T0179670

# THE BIOLOGICAL UNIVERSE

Are we alone in the universe, or are there other life-forms 'out there'? This is one of the most scientifically and philosophically important questions that humanity can ask. Now, in the early 2020s, we are tantalizingly close to an answer. As this book shows, the answer will almost certainly be that life-forms are to be found across the Milky Way and beyond. They will be thinly spread, to be sure. Yet the number of inhabited planets probably runs into the trillions. Some are close enough for us to detect evidence of life by analysing their atmospheres. This evidence may be found within a couple of decades. Its arrival will be momentous. But even before it arrives we can anticipate what life elsewhere will be like by examining the ecology and evolution of life on Earth. This book considers the current state of play in relation to these titanic issues.

**Wallace Arthur** is an evolutionary biologist who is fascinated by the possibility of evolution occurring on other planets. His first book on this subject was *Life through Time and Space* (Harvard 2017), which the Astronomer Royal Sir Arnold Wolfendale described as 'brilliant and thought-provoking in every way'. *The Biological Universe* is the sequel.

'In this thought-provoking book, Arthur's deep knowledge of life and its myriad manifestations, coupled with a cosmologist's understanding of the cosmos at large, enables him to explore one of science's greatest mysteries – how the biological and physical universes relate to one another. Does life exist beyond this planet? What form would it take? How could we detect it? Arthur musters fact, logic, and intuition, in his far-reaching attempt to nail down life's place within the wider cosmic dimension.'

  - Addy Pross, Ben-Gurion University of the Negev, Israel

'Wallace Arthur's book *The Biological Universe* is highly significant. We will soon know whether we are alone in the universe. The next few years could provide us with this long-sought answer. This book, extremely well written, tells us how.'

  - Simon 'Pete' Worden, Executive Director, Breakthrough Initiatives, Luxembourg and USA

# The Biological Universe

## Life in the Milky Way and Beyond

WALLACE ARTHUR
National University of Ireland, Galway

# CAMBRIDGE
## UNIVERSITY PRESS

University Printing House, Cambridge CB2 8BS, United Kingdom

One Liberty Plaza, 20th Floor, New York, NY 10006, USA

477 Williamstown Road, Port Melbourne, VIC 3207, Australia

314–321, 3rd Floor, Plot 3, Splendor Forum, Jasola District Centre, New Delhi – 110025, India

79 Anson Road, #06–04/06, Singapore 079906

Cambridge University Press is part of the University of Cambridge.

It furthers the University's mission by disseminating knowledge in the pursuit of education, learning, and research at the highest international levels of excellence.

www.cambridge.org
Information on this title: www.cambridge.org/9781108836944
DOI: 10.1017/9781108873154

First published 2020

Printed in the United Kingdom by TJ International Ltd, Padstow Cornwall

*A catalogue record for this publication is available from the British Library.*

*Library of Congress Cataloging-in-Publication Data*
Names: Arthur, Wallace, author.
Title: The biological universe : life in the Milky Way and beyond / Wallace Arthur, National University of Ireland, Galway.
Description: Cambridge, United Kingdom ; New York, NY : Cambridge University Press, 2020. | Includes bibliographical references and index.
Identifiers: LCCN 2020020943 (print) | LCCN 2020020944 (ebook) | ISBN 9781108836944 (hardback) | ISBN 9781108873154 (epub)
Subjects: LCSH: Life on other planets. | Exobiology. | Milky Way.
Classification: LCC QB54 .A78 2020 (print) | LCC QB54 (ebook) | DDC 576.8/39—dc23
LC record available at https://lccn.loc.gov/2020020943
LC ebook record available at https://lccn.loc.gov/2020020944

ISBN 978-1-108-83694-4 Hardback

It now appears that most stars host a planetary system. Many of them have a planet similar in size to our own, basking in the 'habitable zone' where the temperature permits liquid water. There are likely billions of Earth-like worlds in our galaxy alone. And with instruments now or soon available, we have a chance of finding out if any of these planets are true Pale Blue Dots – home to water, life, even minds.

Yuri Milner *et al.*,
*launch of the Breakthrough Initiatives, 2015*

# CONTENTS

# PREFACE

The aim of this book is to consider the possible extent and nature of the biological universe, in other words the subset of the physical universe that consists of life-forms. One day, perhaps, we humans can aim to describe the actual biological universe, or at least a sample of it that includes a large number of inhabited planets. But that day is, at best, in the distant future. To achieve the more limited aim that I've set myself here – of 'considering the possible' – the book proceeds through four steps. First, I describe life on Earth, with a focus on the variety of habitats it is found in and the ways in which organisms survive there; and I consider the possible implications of terrestrial habitats and survival strategies for life elsewhere. Second, I examine the current state of knowledge about planetary systems, with a focus on habitability; and among other things I produce an estimate of the number of planets in the Milky Way that are likely to host life. Third, I consider the possible nature of extraterrestrial life, our ongoing search for it, and the link between the two. It's hard enough to find a needle in a haystack, harder still to find a tiny object of unknown shape and composition; and our galaxy is a very large haystack. Finally, I broaden the scope of discussion from its earlier focus on the Milky Way to encompass the universe as a whole.

The structure of the book is based on these four steps. They are taken sequentially in Parts II, III, IV, and V – preceded by Part I, which is an evolutionary and galactic lead-in. Each part is prefaced by a statement of its key hypotheses. The book fits squarely into the genre of popular science, and is written accordingly. Its subject area would best be described as astrobiology, a branch of science that's still in its infancy.

Astrobiology will only come of age when we find our first evidence for extraterrestrial life-forms, something that I expect will happen in the next couple of decades. But infancy is a fascinating stage in the development of a branch of science, just as it is in the development of a person. And this particular infant has a longer history than might be expected: human contemplation of the possibility of extraterrestrial life goes back beyond the Age of Enlightenment to Giordano Bruno in sixteenth-century Rome; probably much further back to ancient Greece, where the heliocentric solar system of Aristarchus pre-dated that of Copernicus by about 1800 years; and possibly further still – perhaps the first members of our species had thoughts about distant life as they contemplated the stars above Africa.

In case you've already read my 2017 book *Life through Time and Space*, or are thinking of reading it, I should point out here both the connections and the distinctions between that book and this one. Both combine astronomy and biology, but they do so in very different ways, and with different foci. A central theme of the earlier book was origins. More than half of its chapters were focused on how things originated; and those 'things' were as diverse as planetary systems, the animal kingdom, and the nervous system of a human embryo. The current book focuses on the present rather than the past. For example, it asks not about how planetary systems arose but what are their prospects for hosting life – in much more detail than the latter issue was dealt with in *Life through Time and Space*. Also, while that earlier book spanned the cosmos right from the start, most of the present one focuses on our home galaxy, the Milky Way, because this is the realm in which we can most easily undertake searches for life.

# ACKNOWLEDGEMENTS

The writing of a mainstream science book covering all the subjects that I touch on here would require a team of authors with the following specialisms: evolutionary biology, ecology, geology, atmospheric science, planetary science, astronomy, astrobiology, astrophysics, cosmology, and philosophy. However, for a *popular science* book, the uniformity of style that a single author can provide is highly desirable. Also desirable is a lack of technical detail, though such an approach carries with it a risk of the treatment being too superficial. One solution to this problem is for the author, who is necessarily a specialist in only one field (in my case evolutionary biology), to read widely in the others before putting proverbial pen to paper, and also to arrange for specialists in the other fields to scrutinize the draft manuscript and provide critical comments. This is the approach I have adopted here. I am enormously grateful to the following friends and colleagues, all of whom have read one or more draft chapters: Chris Arthur, Andy Cherrill, Mark Davies, Mike Guiry, Ernst de Mooij, Dave Newton; and I must say a special thanks to Jim Kasting and Fred Stevenson, both of whom read the whole manuscript and made numerous helpful suggestions for improvements to the text. As is the time-honoured tradition in this situation, I should emphasize that I rather than they am fully responsible for any errors, omissions, or failures of clarity or depth that have somehow escaped the net and filtered through into print.

As well as receiving help from the subject experts listed above, I have also had considerable assistance from publishing and artistic professionals of various hues. At Cambridge University Press, Dominic Lewis was enthusiastic from the moment I first approached him about this project, and guided me through the review process. Sam Fearnley took over from Dominic when things moved from review to production, and efficiently took me through the whole production process. Hugh Brazier cast his inimitable editorial eye over the manuscript and drew my attention to various issues throughout, which together we fixed through an interaction that was at the same time productively serious and great fun. My son Stephen Arthur produced almost all of the final artwork from what can only be described as my amateurish initial scribbles. The only illustration he had no hand in was Figure 18, the Hubble Ultra-Deep Field, which was produced at Cambridge from the NASA original. It's wonderful that all those beautiful NASA photographs of our universe are in the public domain.

*Part I*

# Painting Big Pictures

## Key Hypotheses

### The Terraspermia Hypothesis
Life on Earth originated here, not elsewhere. It's unnecessary to invoke the putative space-travelling spores of the alternative Panspermia hypothesis.

### The Mosaic Hypothesis
The surface of a potentially habitable planet is a mosaic of patches, on each of which life could originate. For any one patch, the probability of origination may be low, but collectively it is high. Having originated in one patch, life can spread to many others.

### The Common Earth Hypothesis
To an order-of-magnitude level of accuracy, there are probably at least a million planets with animal and plant life in the galaxy. The alternative Rare Earth hypothesis is flawed, as we will see.

# 1 A TREE WITH MILLIONS OF TWIGS

## The Tree of Life

The expression 'tree of life' is shorthand for four billion years of birth, death, reproduction, and relatedness. This extended family tree has been produced by four billion years of using energy from the environment to power biological systems. At present these systems, with which our planet is teeming, seem unique in the vastness of the cosmos. But they're not. Their *apparent* uniqueness is an artefact produced by current limitations to human knowledge. One day we will have evidence of life on other planets, and that day may be close at hand. It's not unreasonable to believe that our first evidence of extraterrestrial life will arrive in the next couple of decades.

In this book, our starting point for thinking about life in an interstellar context is the nature of life on Earth. Here on our home planet one particular tree of life has played out. This tree will continue to grow, though the directions in which its still-ungrown branches will extend are impossible to predict, so we cannot look with clarity into our evolutionary future. But we most certainly can examine our evolutionary past. And we can ask to what extent we would expect major features of that past to apply to trees of life that are playing out independently of ours – right now – on planets scattered across the Milky Way galaxy and beyond.

Notice that 'tree of life' is in the singular in the context of our own planet. Every living creature on Earth is related to every other. We humans are not just related to chimps, gorillas, and orang-utans. We are also related to the rest of the animal kingdom and, beyond that, to the trees we climbed as children, the yeast we use to make bread, and the bacteria that line our guts. The branches of the tree of life have no breaks in them. If we made a three-dimensional model tree of this kind, it would be possible to run a finger down from one terminal twig, such as humans, to a particular ancestor in the distant past, and then back up again to any other present-day twig, for example a maple tree.

But what shape should we choose when building our model? In other words, what shape characterizes the overall tree of life on Earth? It has been depicted in many ways since Darwin sketched an evolutionary tree diagram, in the form of lines gradually diverging from each other, in Chapter 4 of *The Origin of Species*. There are several caveats here, because the shape of the tree of life – or of parts of it – has been a source of heated argument among biologists over the years. So we need to tread carefully.

First, scale may be important. Let's consider this in terms of the two-dimensional trees that have been drawn on pieces of paper ever since Darwin. The shape of one small branch and the shape of the overall tree may not be the same. Second, at any scale we choose to examine, the divergence of branches may be leisurely (picture a V) or rapid (picture a U with a flat base). The former corresponds to a 'gradualist' view of evolution, the latter to either a 'punctuationist' or 'saltationist' view depending on the scale. Third, the vertical axis can represent time in an exact way, so that it could be labelled with units such as millions of years; or it could just represent time in a more general way in that it shows only the *order* of branching events, not their relative distances apart. Fourth, the horizontal axis could represent 'degree of difference' or it could be there simply to allow us to picture divergences – something that can't be done unless you have at least a two-dimensional diagram. The difference between these types of horizontal axis is that in the former case the distance apart of two twigs is a measure of their biological disparity, whereas in the latter it is not.

All the above four issues have been the focus of major debates at some stage in the history of evolutionary biology, and some of them continue to be debated. But the purpose of this book is not to examine such issues. We have a bigger picture to paint, so we'll sweep these issues under the proverbial carpet and focus on something even more important – the question of whether a tree diagram of *any* kind is the right way to depict evolutionary relatedness in the first place.

Consider for a moment an actual tree, whether a maple, an ash, or an oak. If you inspect it carefully in winter when no foliage obscures its branches and twigs, what you'll see are thousands of divergences but not a single convergence. Twigs grow apart from each other; they do not grow together and unite. But in the tree of life such growings-together do indeed happen to a degree. Two processes are responsible – interspecies hybridization and horizontal gene transfer. In the former process, two twigs, each representing a single species, hybridize and thus create a descendant species that is different from both of its parents. In theory this shouldn't happen, because a species is defined by its inability to interbreed with others – but in practice it does happen, because definitions are rarely perfect in the biological realm. In the latter process, DNA (deoxyribonucleic acid) from one twig is transferred into another, often via a virus.

The importance of these processes varies according to position in the tree. In the animal kingdom as a whole, their role is minor compared to twig divergence – though that does *not* mean that they aren't important. Some human genes appear to have originated by horizontal transfer from other species, including those as different from us as bacteria. Some of the best examples of interspecies hybridization come from the plant kingdom, while horizontal gene transfer is especially important in microbes.

How should we modify our picture of the tree of life to incorporate these two processes? Hybridization can be included simply by picturing twigs growing together – at least within some of the tree's branches. Horizontal transfer is probably better pictured as a sort of thin wire connecting two twigs at the same level (i.e. the same point in time). Taking both of these modifications on board (Figure 1.1), we now have a tree of life

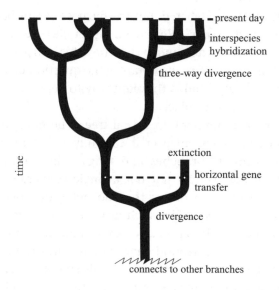

**Figure 1.1** Part of the tree of life on Earth, showing divergences, interspecies hybridization, horizontal gene transfer, and extinction. All these features occur in any reasonably large branch of the tree, though their relative frequencies are expected to vary from one branch to another.

that is still largely tree-like but with some additional forms of growth compared to a real tree. A 2018 book by John Archibald – *The Tangled Tree* – provides further discussion of this issue.

The final thing to say about our tree-of-life picture (or model) is that its top should be flat. It's more like an African *Acacia* tree than a Norway spruce. This is because the present moment of time is the same for all the growing twigs, which collectively represent today's biota – the animals, plants, and other life-forms that populate the Earth right now. Let's take a look at this particular time-slice of the Earth's biological history.

## Present-Day Twigs

So, now we alter our angle of view of the tree from the side to the top. We hover over it as a kestrel might, to achieve the proverbial bird's-eye view that we want. And we look at it as a photographer would when taking one of those shots where the foreground – in this case the present – is in sharp focus, and the background – in this case the past – is just a blur. We are then looking at a series of small circles, each one of them the tip of a growing twig. One is the human circle, another the bonobo circle, and so on. Species of cacti are represented by small circles far away from the ape ones. And mushrooms are represented by small circles far away from both of those other clusters.

Each circle is a species, though as we've already seen species can be badly behaved. The usual definition of a species is that while its members can breed among themselves none of them can breed with members of other species. And there is usually the proviso 'in the wild', so that we exclude information on what can happen in captivity, such as the production of ligers (lion–tiger hybrids). Of course, it would be naïve to expect all real organisms to conform to such a neat human concept. Some do, some don't. But even those that don't can be seen as fitting the definition in a probabilistic way – the density of reproductive interactions among members of a species is much higher than the density of such interactions between them and their sibling species.

Because there are at least a few million species on the Earth at present, and perhaps a few tens of millions, we need to have some way of structuring our knowledge of this vast biodiversity. And what better way than the method provided by the Swedish naturalist Carl Linnaeus in the mid-eighteenth century. Taking his approach, we group a bunch of neighbouring twigs together by drawing larger dotted circles around their small solid circles, thus representing groups of related species called genera (singular genus). For example, the orang-utan genus (*Pongo*) includes three twigs – those of the Bornean, Sumatran, and Tapanuli orangs (Figure 1.2). Our own genus (*Homo*) consists of only a single species in today's fauna. In contrast, some genera – for example the insect genus *Drosophila* – have hundreds of species.

In this exercise of looking down from above on the growing tips of the tree of life's twigs and drawing circles, we are doing something that can be described in terms of set theory. Our

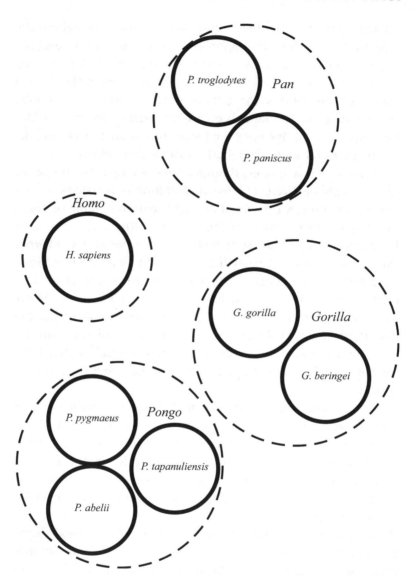

**Figure 1.2** A particular part of the tree of life – the great ape branch – seen from above. Note the single extant species of humans (*Homo sapiens*), in contrast to the two or three species each of chimps, gorillas, and orangs. Common names are: robust or common chimpanzee (*Pan troglodytes*); pygmy chimpanzee or bonobo (*P. paniscus*); western gorilla (*Gorilla gorilla*); eastern gorilla, including mountain gorilla (*G. beringei*); Bornean orang (*Pongo pygmaeus*); Sumatran orang (*P. abelii*); Tapanuli orang (*P. tapanuliensis*).

circles-within-circles picture is what a mathematician would describe as a large set that includes one or more smaller sets. But there's something unique about our taxonomic sets: they are related to each other by their shared branches of the past. In a set of crockery types, where one subset is 'cups', another 'plates', and so on, there is no such underlying common ancestry – each item is made from scratch.

Taxonomists sometimes describe what they're doing as discovering and describing 'the pattern of natural classification'. The Linnaean approach draws bigger and bigger circles around progressively greater numbers of twigs, so that after species and genera we have families, orders, classes, and so on. Not only are these progressively more inclusive in terms of current biodiversity, but they are also progressively more deeply rooted in the tree of life. At the most inclusive end of the taxonomic hierarchy in Linnaeus's scheme was the kingdom – still in use today but expanded in number. Linnaeus described just two kingdoms of life – plants and animals. Now we also recognize at least one more – the fungi – and almost all biologists would say that there are several others. For example, all the large conspicuous brown seaweeds that we observe around our coasts, including those that make up that wonderful marine habitat called the kelp forest, are outside of the plant, fungal, and animal kingdoms. Studies on their genes make this conclusion clear. Collectively they are brown algae, but beware the term 'algae' as it has many inconsistent usages. They are in a fourth kingdom, even though there is some debate over what its name should be.

For Linnaeus, above kingdoms of life there was simply 'life'. But now we insert an even higher level of taxon than kingdoms – domains. The American microbiologist Carl Woese refined the taxonomic scheme of the earlier Carl in a major paper published in 1990. He grouped life-forms on Earth into the domains Bacteria, Archaea, and Eukarya. The first of these is self-explanatory, the last contains animals, plants, fungi, brown algae, and all other life-forms that are built of complex (eukaryotic) cells. The middle one, Archaea, was new, and based on earlier work by Woese and his colleagues. Superficially, the organisms that comprise Archaea look like bacteria, and pre-Woese they'd been classified as such. But, as he showed, they have a different form of RNA (ribonucleic acid) from the other two domains; and they use different fats in their cell membranes too. These are very deep-seated differences, and reflect their early divergence from both bacteria and eukaryotes.

It's important to realize that all the taxonomic categories above species – from genera up to domains – are arbitrary and have no clear definitions. They simply constitute a useful way of organizing information. The species is the only category that has biological meaning for the entities that comprise it, as opposed to for human observers; hence our ability to define, albeit imperfectly, what 'species' means. But there's an even more fundamental definition that we now need to consider – that of life itself.

## What is Life?

If you'd like a lengthy discussion of this issue I can recommend the 2012 book of the same title by the organic chemist Addy Pross. Here we'll focus on just two approaches, which I'll call evolutionary and metabolic. The first is tightly linked to Darwinian natural selection. The second is linked instead to the biochemical processes that go on within cells. It's quite possible to be alive by one definition but not by the other; indeed, that's the case with viruses.

The evolutionary definition of life is as follows. Entities that exhibit the three properties of variation, reproduction, and inheritance are alive; those that don't are not. These are the very same three properties that are necessary for natural selection to occur. Consider a group of entities – we'll not prejudge the issue by calling them organisms just yet – that are rather similar but not identical to each other. They reproduce, in at least one of an immense variety of ways (beautifully discussed by Italian biologists Giuseppe Fusco and Alessandro Minelli in their 2019 book *The Biology of Reproduction*), and the offspring resemble their parent(s) more than they resemble randomly chosen members of the group. Resemblance in this context is not just external, nor just structural; it is internal and behavioural too. In such a situation, whichever variants are best suited to the current environmental conditions will leave most offspring, and so the composite nature of the population will change over time. Such a situation does not 'give rise to' natural selection – rather, it *is* natural selection.

At first this definition seems clear. According to it, birds and ferns are alive, while rocks and clouds are not. But if we dig deeper we find problems. Mules seem just as alive to me as do the horses and donkeys that were their parents. But, as sterile hybrids, they generally cannot themselves reproduce. Surely we shouldn't leap to the conclusion that they are inert entities, non-life-forms. And the converse problem of an entity that has the three requisite properties to be considered life but that we generally do not think of as life can also be encountered – for example computer viruses. These can exhibit variation, reproduction, and inheritance, but most of us would not consider them to be alive. And what about real (biological) viruses – are these alive? Many biologists see them as inhabiting a philosophical grey area between the living and the non-living. They can reproduce, but not on their own without hijacking another living system to help them. Then again, the same could be said of a tapeworm. So the evolutionary definition on its own is problematic.

The metabolic definition of life goes something like this. An entity is alive if it takes up energy and materials from its environment, uses these to maintain an internal state that is dynamic and yet buffered to some extent from environmental fluctuations, and ejects waste products from this process back into the environment from which the raw materials came. For the most part, this definition classifies entities as alive or not in the same way as does the evolutionary one: birds and ferns are alive, rocks and clouds aren't.

But again problems emerge when we start digging. The tiny invertebrate animals called tardigrades (or water bears) are famous for being able to withstand extreme conditions. They can survive extended periods of temperatures close to absolute zero, which would freeze-kill most other animals very quickly. They go into a state of suspended animation, from which they wake up when the ambient temperature is increased again. They use a similar technique to survive the vacuum of space; some of the tardigrades that have been taken into space – on the outside of a spacecraft rather than in the relative comfort of its interior – have survived and reanimated themselves on return to Earth, as reported by the Swedish scientist Ingemar Jönnson and his colleagues in 2008. Is a 'cryptobiotic' tardigrade alive? Personally I'd say yes, and that it's just a rather extreme form of a hibernating hedgehog; but not everyone will agree with this view.

And what does being 'buffered to some extent' mean? We mammals can maintain an internal body temperature buffered into a narrow range around 37 degrees Celsius (98 Fahrenheit). Crocodiles can't do that. Their internal temperature is much more variable over time – though it's still buffered 'a bit' from the prevailing temperature of the environment. This more modest buffering is partly metabolic and partly behavioural in the sense of a crocodile's choice of microhabitat.

What about the converse problem to that presented by the deep-sleeping tardigrade, in other words an entity that could be called 'alive' by the metabolic definition but which common sense would suggest is not? A fridge takes up energy from outside itself, uses this to maintain a regulated internal state, and ejects heat back into the environment as a sort of waste product. In this case we perhaps escape from definitional problems in that the fridge doesn't take up *materials* from its environment. But even then a qualification is needed, because it does take up materials (cartons of milk, bottles of beer) – but not without human help, and it doesn't use them to produce its internal homeostasis.

The metabolic definition could be modified by adding a stipulation that the 'inside' and 'outside' of the entity we're looking at should be separated by one or more membranes – otherwise we conclude that the entity is inert. Again this works to distinguish birds and ferns from rocks and clouds. But is it too restrictive? Its use would remove viruses from their grey area and classify them as non-living, which to me seems too certain a conclusion for such hard-to-classify entities. And if we want to have a definition of life that we can apply to entities that we find on other planets, then it might be unwise to insist that we definitely won't call them life-forms if they don't have membranes, especially in our current state of ignorance as to what form life takes elsewhere than on our own planet.

So, how to proceed? I suggest that we adopt the following policy. If an entity is metabolically alive and membrane-bound, and groups of individual entities of this kind are characterized by variation, reproduction, and inheritance, then we describe the situation as 'life'. If none of these criteria are met, the type of entity we're examining is inert – i.e. not alive. Situations in between these two, such as viruses, we treat on a case-by-case basis. And regarding extraterrestrial life we should try to keep as open a mind as possible – we'll revisit this issue in Chapter 3.

## The Omnipresent Force

A common distinction made by biologists about evolution is between pattern and process. Evolutionary pattern is what is produced after long-term operation of the process. The tree of life is a pattern, but although its existence is good evidence for some kind of evolutionary process having occurred, it tells us little about the nature of the evolutionary process or the mechanisms underlying it. What forces have *driven* it? More than one, for sure, but what I'm calling here 'the omnipresent force' is Darwinian natural selection. In some sense, this is the most important force, though that seemingly simple statement is trickier than it initially appears.

It's easiest to observe selection in action when the process happens quickly. And for this the best combination is a major threat to life (i.e. strong selection) and a short generation time. Hence one of the best examples of the power of selection is the evolution of antibiotic resistance in bacteria, now a major public health concern. In this situation, bacteria are facing a novel chemical designed to exterminate them. In a particular population of bacteria exposed for the first time to a new antibiotic, the mortality level may well be in excess of 99%. But, due to natural variation among individuals, a few may have inbuilt resistance to the threat, just because of some tiny difference in metabolism compared to all their relatives. In such a situation, the progeny of the resistant individuals will prevail. If many generations can unfold in hours or days rather than years or decades, evolution of the population to the point where it is 99% resistant as opposed to 99% susceptible will be very rapid.

Evolution of resistance to pesticides among insects is a similar phenomenon except that the typical insect generation time is much longer than the typical bacterial one, so the process takes a significantly extended period. Evolution of grasses and other plants to growing on the metal-contaminated spoil heaps surrounding old mine workings is similar too; as is the familiar textbook example of the evolution of certain insect species in response to smoke pollution caused by the Industrial Revolution. In the best-studied instance of this – the peppered moth – evolution from 99% peppered (camouflaged against the old lichen-covered clean tree trunks and branches) to 99% quasi-black (camouflaged against the novel sooty tree trunks and branches) took about a century.

Whether such examples constitute 'natural' selection is a moot point. It depends on the extent to which human-created agents of mortality are considered to be natural. But there is no clean line separating these anthropogenic environmental threats from others. Sometimes rapid environmental change can cause very strong selection without any human interference. It seems likely that this was the case when the asteroid that killed off the dinosaurs (and more than 75% of species of everything else too) crashed onto the piece of land that is now called Mexico. Although for most creatures mortality was complete, for the lucky few it was merely 'very high'. The border between extinction and rapid evolution is a narrow one.

Most of the time, over most of the world, selection probably works quite slowly. But that doesn't make it any less important – just harder to catch happening. When there are many geographical boundaries between small areas, in each of which selection favours different forms, we have an arena in which selection may be studied on what we might call a middling timescale – neither months nor millennia. Archipelagos constitute just such arenas, with the evolution of Darwin's finches on the Galapagos archipelago being the best-known example. An early classic book on this system by British ornithologist David Lack has been supplemented by more recent books written by Princeton-based biologists Peter and Rosemary Grant.

Demonstrating natural selection in action as a driving force of evolution is one thing; demonstrating Darwin's 'theory of natural selection' is quite another. In fact, we now need to ask: what exactly is this theory? To me, it is best summed up by Darwin's own words at the end of the Introduction to *The Origin of Species*: 'I am convinced that natural selection has been the main but not exclusive means of modification.' How do we demonstrate not just that natural selection is *one of* the forces driving evolution but that it is *the main* one? Clearly, we must specify other such forces and somehow assess their relative importance. And we must acknowledge that what has been included in the 'other' category has changed since Darwin's time. Now, in the early twenty-first century, the other forces of which we need to take account are: random genetic drift, gene mutation, and developmental channelling. Of these three, only genetic drift is a direct challenger to natural selection; the others are complementary to it, as we'll see below.

Genetic drift and natural selection are both things that influence *populations* of organisms as opposed to individual organisms. In that respect, they're similar. Where they differ is that natural selection is a systematic process producing a quasi-predictable result, at least in the short term, while genetic drift is a random process whose influence is indeterminate. But describing drift as a random process is not sufficient to give a clear picture of the nature of the beast. So let's zoom in on it more closely.

A digression into coin-tossing is helpful at this point. If I toss a coin, the odds of it ending up heads are 50%. If I toss it five times, the odds of getting five heads are less than 10%. So usually a person tossing a coin five times doesn't get this result. But if you have a large lecture theatre full of students and you ask them all to do this experiment simultaneously, some of them *will* get the magical five heads.

The same applies to the fate of variant versions of genes in populations of organisms. Suppose there is a gene called $g$, and in a particular population of animals there are two versions of it – $g_1$ and $g_2$. If one of them – say $g_2$ – conveys enhanced fitness, it will tend to spread through the population by natural selection. However, if $g_1$ and $g_2$ don't affect fitness, their relative frequency in a population will be determined by genetic drift. This is like coin-tossing. The probability of the frequency of $g_2$ going up rather than down in one generation is 50% (assuming that its chances of remaining *exactly* the same are negligible). The probability of it going up in five successive generations is less than 10%. Yet if we have enough populations of the species concerned scattered across the globe, some of them will show exactly this outcome. And in a few of them $g_1$ may be lost entirely, with $g_2$ thus becoming 'fixed' in the population, despite the lack of natural selection.

In the early days of population genetics, this result of genetic drift was thought by most biologists to apply only to small populations. But the Japanese geneticist Motoo Kimura showed that drift could be important in large populations too. He devised 'the neutral theory of molecular evolution' (and published a book of the same name in 1983), according to which, at the level of macromolecules (DNA, RNA, and proteins), genetic drift may be responsible for *most* of the changes that we observe in natural populations. If this is true, Darwin's theory of natural selection, at least in the form I've given it here, is wrong. Interestingly, although Kimura promoted this view, he also argued that Darwin's theory was correct at the level of the whole organism. For example, he agreed with Darwin that giraffes evolved long necks because of natural selection, not because of genetic drift.

It seems paradoxical that drift could be the main agent of change at the molecular level, while at the same time natural selection is the main agent of change at the level of the whole animal (or plant or microbe). But actually it's fine. If the vast majority of single amino-acid changes in a protein that is made up of 500 of them don't affect its function, a predominance of drift at this level might be expected. But by definition none of those changes will affect the animal's form. When we focus on morphology these neutral changes are invisible. The only molecular changes that are relevant to neck length in giraffes, or to body size and shape in animals more generally, are: (a) those that occur in genes that have an effect on body size/shape; and (b) those that affect the functioning of such genes and their products. These changes will all be subject to natural selection.

In my view it remains to be seen whether Kimura was right about the predominance of drift at the molecular level of organization. But regardless of this, Darwin's belief that natural selection is the 'main' agent responsible for evolutionary change at the organismic level has been borne out by more than 150 years of observational and experimental evidence. Natural selection really is omnipresent in biological systems – providing that there is some variation on which it can act.

## Origins of Variation

We tend to take biological variation for granted. It's all around us. And the role that genes play in its generation can be seen from the natural experiment of identical twins. Although such twins are never truly identical, their differences in physical form are amazingly slight compared with those between two randomly chosen humans whose genetic make-ups are different. In humans and most other animals, a major player in reassorting genes and hence in influencing the nature of variation is sexual reproduction. But this is indeed just an influence – albeit a very large one. It only works when there are variant genes present in the first place. Furthermore, the very earliest life-forms probably didn't have sex – at least as we know it. This is still the case with some microbes (and indeed some larger organisms) today. Ultimately, different versions of genes come from the type of molecular accident that we call mutation.

Mutation is a change in the sequence of the DNA that's found in the genomes of all Earthly life-forms (excluding the RNA viruses, if we deem viruses to be alive). A typical gene has more than 1000 DNA bases arranged in a particular order. Mutation involves an accidental change in at least one of them – an accident that usually occurs when the DNA is being replicated. The fact that such changes occur at a background level in all biological systems is hardly a surprise. Rather, the surprise is that the background level is so low – think of the level of errors when someone transcribes a substantial body of text by hand. The rate of mutations can be radically increased by many factors, including exposure to carcinogenic chemicals or ionizing radiation such as x-rays. But such high rates aren't necessary for natural selection to work – with only the background level it works just fine.

However important mutation is, we have to remember that it's just a change in a gene. The step from there to changing the organism as a whole is complex – just how complex depends on the identity of the organism concerned. The step is longer for a mammal than it is for a bacterium. And in any multicellular creature, whether mammal, butterfly, or birch tree, it involves the developmental process. Exactly how this process channels mutational changes, and hence interacts with natural selection, is a fascinating, but still largely open, question – it was the subject of my 2004 book *Biased Embryos and Evolution*.

So, inherited variation is ultimately produced by mutation. The form that it takes is influenced (in different ways) by sexual reproduction and by the developmental system – in the case of organisms that have these things. Natural selection acts on the array of variants at its disposal and moves the average form of the population in a particular direction – by definition the one of higher fitness under the prevailing environmental conditions. Although there are many complexities glossed over here, this is the essence of the main force driving evolution. And it has been in the driver's seat ever since life began. But it cannot in itself explain how life originated. So now we ask: how did that happen?

## Panspermia is Not the Answer

The steps from carbon atoms to life-forms are as follows: first, synthesis of very simple organic molecules such as methane; second, synthesis of more complex ones involving two or more carbon atoms such as alcohols; third, synthesis of the complex organic molecules that are found as repeating units in the macromolecules of life, including amino acids (found in proteins) and sugars (found in nucleic acids and carbohydrates); fourth, formation of the macromolecules themselves; fifth, formation of aggregations and interactions of such molecules together with smaller ones, thus giving rise to a sort of proto-metabolism; sixth, the development of quasi-autonomy from the environment through becoming membrane-bound; finally, the origin of a simple form of reproduction, perhaps involving the budding of smaller membrane-bound units from a larger parental one.

Of these seven steps, the first three are relatively 'easy' and occur all over the universe. They do not require a finely tuned planetary environment. Methane, alcohols, sugars, and amino acids have all been found in space. They arrive on Earth via meteorites, most of which have come from the asteroid belt. The making of these molecules is chemistry, not biology.

The last four steps still involve chemistry, of course, as do all processes involving matter of any kind – with the proviso that we can as yet only speculate about the nature of dark matter. But they also involve biology. Neither DNA nor proteins have ever been found beyond the Earth. Finding an amino-acid molecule in a meteorite that arrives on Earth tomorrow would not be news – except to the tabloid pseudo-press and its electronic equivalent.

But finding a meteorite-transported protein molecule whose tangled form was made up of 100+ amino acids joined end-to-end would be mind-boggling.

The Panspermia hypothesis proposes that terrestrial life began with dormant spores arriving on the proto-Earth from space. Their dormancy was broken by their arrival here and they proceeded to produce the tree of life that was our focus of attention in the first section of this chapter. This hypothesis has been championed by various scientists, including the Swedish chemist Svante Arrhenius and the British astronomers Fred Hoyle and Chandra Wickramasinghe. My personal view is that it is wrong, and that the alternative hypothesis of steps 4 to 7 having occurred on Earth – let's call it Terraspermia – is correct.

My belief in the correctness of the Terraspermia hypothesis is not based on us having a good understanding of the origin of life on Earth – we don't. Rather, it is based on three things: the likelihood that there is no life on Mars, the moons of Jupiter and Saturn, or anywhere else in our solar system; the improbability of the survival of *any* form of life, dormant or otherwise, through the nightmare environment of interstellar space; and the desirability of using Occam's razor. Let's quickly look at these three bases for terraspermia.

Life beyond the Earth but within our solar system can't yet be ruled out, but we have no evidence for it, and at least in some places – such as parts of the surface of Mars – we've looked quite thoroughly. To date, we haven't drilled through the icy surface of the moons Europa (orbiting Jupiter) or Enceladus (orbiting Saturn), so there might yet be life in their sub-surface oceans. But I doubt it. Perhaps some terrestrial extremophiles could survive in such an environment, at least for a while. But could life actually originate and evolve there? That would be much harder.

If there's no extraterrestrial life in our solar system, then an incoming spore would have to come from the planet now known to orbit the nearest star to the Sun – Proxima Centauri – or from even further afield. This means that it would have to survive distances vastly greater than those separating the planets within a single system. The probability of such survival is, in my view, zero. Even those ultimate survival machines the tardigrades, which we met earlier, mostly died in Earth orbit – a fact that is less often emphasized than the survival of a few of them. The difference in journey-time between a few Earth orbits and making the trip from the Proxima Centauri planetary system to our own is immense: the latter would take about a billion times as long, travelling at a similar speed.

It might be better to say that the probability of a dormant spore surviving such a journey would be vanishingly small rather than zero. So there is still a glimmer of hope for panspermia. But this is where Occam's razor comes in. Why propose a hypothesis involving two improbable events instead of just one? Science prefers simpler solutions whenever possible. Amino acids, sugars, and other small molecules becoming proto-cells containing macromolecules of DNA and proteins is a tall order anywhere; it doesn't become any easier if we transplant it to a planet orbiting Proxima Centauri. But if we do so transplant it, we must invoke a second tall order – that of surviving a journey of more than 4 light years. So terraspermia is preferable to panspermia on the basis that it requires fewer improbable causes.

However improbable is an origin of life from routinely found organic molecules on any particular initially barren planet in any particular millennium of its history, the probability of life originating rises massively when we consider vast collections of planets over vast spans of time. And that takes us to where we're going next – the entire Milky Way galaxy.

# 2   A GALAXY WITH BILLIONS OF STARS

## Portrait of the Milky Way

We need to have a clear mental picture of this vast galaxy of ours, the Milky Way, in order to proceed with considerations about the life it may contain. But it's hard for any artist to paint a portrait of something from within. Our situation in this respect is akin to Leonardo da Vinci trying to paint the Mona Lisa from inside her head. The best way to deal with this galactic difficulty is to start with what we see from the inside, then examine what we think the galaxy looks like from the outside, and then try to understand the connection between the two.

We've all seen the Milky Way as a whitish streak across a dark sky either first-hand or via a photograph. From such a view, it's easy to see how it gets its name. And in languages other than English the name is often similar. The Italian Via Lattea translates as Milky Way, as does the Dutch Melkweg. There are some interesting variants such as the Silver River (Chinese) and the River of Heaven (Japanese). But nearly all refer to a way, road, path, or river, thus capturing the linear nature of the thing we see in the sky.

However, despite the multilingual consensus, this apparent linearity isn't real. All the names derive from ancient times, when the only view to be had of our galaxy was the one that could be seen overhead with the naked eye on a cloudless, light-pollution-free night. But since we've been viewing the heavens with scientific instruments, an era that started with Galileo and his pioneering refractor telescopes, we've been able to get a different view. When Galileo looked at the Milky Way telescopically, he saw that the milkiness resolved itself into numerous stars. This was the first big step forward in human understanding of our galaxy. As telescopes improved we learned more and more. And trans-navigation of the world by European explorers helped too. At about the same time as Galileo first looked at the Milky Way with his telescope in the early years of the seventeenth century, a Dutch expedition arrived in Australia. From such a vantage point in the southern hemisphere our galaxy looks somewhat different. The 'Way' has a thickening, which we now know represents the galaxy's central bulge.

But let's not get ahead of ourselves. Where has this 'bulge' come from? What is it? It's now time to turn to an external view to make sense of the shape of the Milky Way and the nature of its bulge. However, we can't obtain an external view directly – that would involve travelling far enough from Earth to be able to turn round and take a photo of the whole galaxy. Such a journey would take about a million years, even at speeds approaching that of light, which are far beyond our current technological capability. One way around this problem is to look at other galaxies that we think are broadly similar in shape to our own – for example our nearest neighbour Andromeda, or the marvellously named Pinwheel and Whirlpool galaxies.

The above-mentioned three, and our own Milky Way, are all *spiral* galaxies. What this means is that they each have the shape of a mutant Frisbee. To get a spiral-galaxy appearance from a Frisbee you need to do four things: first, cut off the downward-pointing rim; second, swell the plastic in the centre into a bulge; third, paint on a pattern of spiral stripes, like those arcs of fire produced by a Catherine wheel; fourth, embed the mutant Frisbee in a roughly spherical micro-cloud.

With this Frisbee-inspired mental image we can now name the parts of the galaxy and say a little about each. The central bulge – a very dense collection of stars – gives way, going outward, to the disc – the name for the flat part. The disc is characterized by bright curved arms, areas that stand out visually from the rest of the disc because they have many young, bright stars. We'll look at these arms in more detail shortly.

The cloud is the stellar halo – a region of sparsely distributed and mostly old stars.

The above is a very stellar-centric view. Galaxies also contain a lot of gas and dust. Indeed, a view of the Milky Way arcing across the night sky shows not just the milky streaks of stars but also dark patches where dust obscures some fields of stars from our view. And then there's the mysterious dark matter whose existence we infer from indirect evidence – such as the relative rotation speeds of the Milky Way's many stars – but which we can never directly see, since it doesn't emit or reflect any light.

The picture I've painted so far is somewhat static. Only the brief reference to rotation and the use of a Catherine-wheel analogy have hinted at movement. But in fact galaxies are very mobile things. Spiral galaxies are constantly rotating, for example. And each galaxy moves relative to others. From time to time this leads to galactic mergers. The Sagittarius dwarf galaxy is in the process of a merger with the Milky Way, and Andromeda will merge with us in the distant future. When two spiral galaxies merge – as in the latter example – the result is often a giant elliptical galaxy.

Now let's connect our external and internal pictures of the Milky Way. From our home location part-way out from the central bulge and at an unremarkable position on the disc, we look into a dark sky, and what we see depends on the direction in which we look. When we look into the plane of the disc we see a band of dense stars – the 'visual Milky Way'. When we look at another angle we see the sparser stars above and below this plane – stars that to a non-informed observer are outside the Milky Way but in fact are within it. If we look up at the sky from Wales we never see the central bulge; but if we look up from New South Wales we always see it, at least when the sky is clear enough. The reason for this at-first-sight odd situation is that the plane of our solar system is closer to being at right angles to that of the galaxy than parallel to it (Figure 2.1). Earth's northern hemisphere permanently points outward, in galactic terms, while the south points inward. The Earth's rotation on its own axis, its axial tilt, and its orbit around the Sun do not alter this

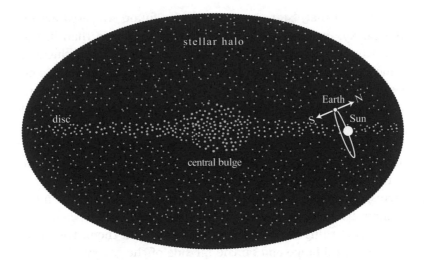

**Figure 2.1** Side-on view of the Milky Way, showing the galactic central bulge, the disc, and the stellar halo. The solar system's location and orientation are also shown, but in simplified (Sun–Earth only) and magnified (millions of times) form. This shows why the central bulge is visible from Earth's southern, but not northern, hemisphere.

arrangement of 'north = outward', which thus prevails throughout the day, and throughout the year.

## How Many Suns?

It's time to be a bit more quantitative. As every artist knows, a good portrait not only needs to be artistic, but also needs to have its proportions right. A beautiful skin tone on a hexagonal face would not be a winning combination – except, perhaps, in the case of some mutant form of Cubism. So, let's put some dimensions onto our picture of the Milky Way. To do this, we need to use the units of distance called light years. One light year represents about 10 trillion kilometres or 6 trillion miles – the distance that light travels through space in a year.

Our galaxy is about 100,000 light years across, but its disc is 'only' about 1000 light years thick. Stars are most tightly packed in the bulge, less so in the disc, and very much less so in the halo.

Having said that, 'star clusters' are also found scattered across the galaxy; in these, stars are more tightly packed than in the inter-cluster surroundings. They are of two types: open clusters, which consist of young stars, and globular clusters, which consist of old ones. The open clusters are particularly common in the spiral arms and other parts of the disc. In contrast, the globulars are scattered all over the place, with many being in the stellar halo. But how many stars are there in the Milky Way as a whole? That's the bit of quantification to which we now turn; and it's a particularly important one in astrobiological terms. Planets are the most likely venues for life; and most planets are to be found orbiting stars.

The first thing to note is that we can't count them. For those in the central bulge and on the far side of the galaxy, there are too many foreground stars obscuring them. This is even true of the further stars within our own part of the disc. So we have to find some other way to estimate the number of suns (i.e. stars) in the Milky Way. And here's how it can be done. For a rotating system, such as a spiral galaxy, the mass can be worked out from the speed of rotation. For example, our Sun (and solar system) rotates around the galactic centre at a speed of about 200 kilometres (125 miles) per second. This means that it completes a circuit in about 250 million years. Interestingly, the last time we were at our current galactic position was the time of the 'great dying', the biggest extinction event in the history of the animal kingdom, which eliminated those wonderful armoured marine animals, the trilobites. This was before the (smaller) extinction event that killed off the dinosaurs – in fact it occurred before the dinosaurs had even originated. So in the time it takes for our solar system to do a lap of the galaxy, a lot can happen on Earth in evolutionary terms.

Anyhow, trilobites and dinosaurs aside, from our speed of orbiting the galaxy we can determine the approximate mass of the material within this galactic orbit. And from that and the galaxy's dimensions, we can have a stab at estimating the overall mass. However, one of the lines of evidence for dark matter is that stars towards the edge of the galaxy orbit 'too quickly'. It's now thought that a galactic dark-matter halo extends out further than the stellar halo; and that the amount of dark matter both in the Milky Way and in the universe as a whole dwarfs the 'normal matter' (stars, planets, etc.) with which we are much more familiar.

Putting the problem of dark matter to one side, stars make up the vast majority of the mass of normal matter in the galaxy. And this mass has been estimated as being about 100 billion times the mass of the Sun. If the Sun were an average star, then this would suggest that the Milky Way has about 100 billion stars. But we know that the Sun isn't average. The range of star sizes goes from more than 10 times the mass of the Sun (giants) to less than 10% of its mass (dwarfs), but the distribution is very asymmetrical. The commonest type of star is the diminutive red dwarf. Sun-sized stars are rarer, and very large stars rarer still. Given this skewed distribution of stellar masses, the total number of stars in the Milky Way must be at least 100 billion. Estimates vary from this minimum figure up to about 500 billion.

## The Disc and its Spiral Arms

The disc of our galaxy is especially important in terms of looking for life: first, because all our nearest stars are disc stars; and second, because for various reasons, habitable planets may be more likely to exist in the disc than in the bulge or the halo. So let's probe further into this key galactic component.

The first thing to be said is that, as elsewhere in astronomy, imperfection rules. Humanity started with a view of perfection in the heavens. After Copernicus, our view was of perfectly spherical planets in perfectly circular orbits around perfectly spherical suns. But we've known this picture to be wrong ever since the German astronomer Johannes Kepler demonstrated in the early seventeenth century that orbits are elliptical. And wherever we look in detail at apparent perfection, we find imperfection instead. With regard to the disc of the Milky Way and its spiral arms, we don't even need to look too hard. The disc is warped. And, unlike any disc we're familiar with from day-to-day encounters, such as a Frisbee, saucer, or DVD, it has no clear upper or lower surfaces, and no clear outer edge. Stars simply peter out in all these three directions – up, down, and outward. The Milky Way has a fuzzy outline, not a precise one.

When it comes to the spiral arms, the imperfection is even more pronounced. To a mathematician, there are various perfect spirals, notably the one called the equiangular (or logarithmic) spiral, which is the one to which many snail shells approximate, as shown by the Scottish polymath D'Arcy Thompson in his epic 1917 book *On Growth and Form*. But unlike snail shells, galactic arms don't even approximate any of the perfect mathematical spirals. They're more like what you find when someone has accidentally trodden on a snail shell – the arms are fragmentary and irregular. Inside us, in galactic terms, is the Sagittarius arm, outside us the Perseus arm. Our own arm – Orion – is sometimes called a spur because it's more like an arm fragment than a complete arm.

Here's another dimension to the imperfection of the disc and arms. As we've already seen, stars in the disc orbit the centre of the galaxy at measurable speeds. And the arms rotate around the galactic centre too. However, not only do stars at different distances from the centre rotate at different rates, but also the arms rotate at different speeds than their constituent stars. In the vicinity of our Sun, the stars move more quickly than the arms. This means that, in the very long term, stars at this distance out from the galactic centre approach an arm from behind, move forward into it, remain within it for as long as the relative speeds dictate, and then exit from the front, rushing on ahead of the arm that for a long while was their home.

One theory of the arms is that they are caused by density waves sweeping around the disc. High density causes clouds of gas and dust to collapse under their own gravity more readily than they would otherwise have done. Such density-induced collapse means that the arms are sites of enhanced star formation. And this is what makes them stand out from the rest of the disc.

## Meditation on Orion

Imagine this: you stand outside on a clear night at the top of a hill somewhere far from any major city. You look up at the sky – choosing a particular direction in which to do so, for example south. You pretend that a huge patch of southern sky is enclosed in a rectangle of wood, a sort of giant picture frame. Let's suppose that this frame includes the constellation of Orion the Hunter, probably the best-known constellation after Ursa Major (the Great Bear, with its familiar 'big dipper' or 'plough'). The main features of the Orion constellation (Figure 2.2) are the three stars of the famous belt, the two bright stars above the belt (Betelgeuse and Bellatrix), and the two below (Saiph and Rigel). There's also the fuzzy 'star' (not a star at all) about halfway down the sword that dangles from the belt – this is the Orion Nebula star-forming region.

The brightest of these stars are Betelgeuse (top left, sometimes called Beetlejuice) and Rigel (bottom right). If you look at these two stars you'll see a pronounced difference in colour: Betelgeuse is reddish, Rigel bluish. This is due to a difference in surface temperature, with Rigel being much the hotter of the two – our own Sun is hotter than Betelgeuse but cooler than Rigel.

Although we can infer their temperatures from their colours, there is no way we can infer their distances simply by looking at them. Since they're both very bright we might be tempted to think that they're equally close – but they're not. Betelgeuse is about 650 light years away, Rigel about 850. This is just one particular example of a more general phenomenon. Of the three almost equally bright stars of the Summer Triangle, two of them (Vega and Altair) are about 20 light years away, the other (Deneb) over 2000. The crucial point is this: the night-time sky appears flat. We can see two dimensions, up–down and side-to-side, but we cannot see the third, the in–out dimension. The apparent brightness of a star doesn't help in this respect because there is a trade-off between intrinsic brightness and distance: a bright distant star and a dim close star can look identical.

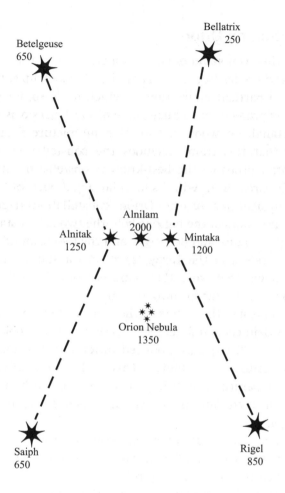

**Figure 2.2** The seven main stars of the Orion constellation, shown with their approximate distances from Earth, measured in light years. Also shown is the Orion Nebula star-forming region. Notice that the distance between the nearest and furthest of these stars is about 1750 light years – many times further than the distance from Earth to the closest of Orion's stars.

What this means is that constellations do not exist in the way that, for example, individual stars and star clusters exist. A short distance away from the top right of Orion (in a northern-hemisphere view) is the open star cluster called the Pleiades or Seven Sisters. Although the number of stars in the cluster appears to be sevenish to the naked eye, there are actually more than 1000 of them. And they really do make up a physical, three-dimensional cluster of stars in space, which is about 400–500 light years away. They're all about the same age – in fact they're all relatively young stars that were probably all born from the same collapsing giant molecular cloud complex.

If we had a sci-fi spacecraft with warp-drive, we could depart from Earth, fly the 400 light years or so, circle the cluster, and return home with the photos to prove it. But we couldn't do that for the Orion constellation. We could try, of course. But what would we find if we did try? Suppose that we travelled the 650 light years to Betelgeuse and paused to take stock of our new position. The middle star of the belt – Alnilam – would still be well over 1000 light years ahead. This means that the Earth is much closer to Betelgeuse than is Alnilam. Does this mean that our own Sun is part of the Orion constellation? The answer is an emphatic 'no'.

This thought exercise illustrates very clearly that while some groupings of stars that we can see from Earth are real – such as the Seven Sisters – others are not. And in 'others' we have all the 88 official constellations. These are human constructs derived from searching for recognizable patterns of stars from our particular location in the galaxy, and with the restriction of not being able to see that in–out dimension of the sky. As soon as this dimension is added, our constellations become meaningless. Luckily, this does not detract from their visual beauty. Indeed, it enhances such beauty in a quasi-philosophical way. I often gaze at Orion, picture the ancient mythological hunter after whom the constellation is named, and consider what he might look like from other angles. The idea of a belt that is hundreds of light years away from the figure that is supposedly wearing it is perhaps an ideal way to recall the artificial, human-centred (in more ways than one in this case) nature of the things we call constellations.

But 'Orion' isn't just a constellation. It's also, as we've already seen, the name of one of the Milky Way's spiral arms. And the arms are real enough as regions of high stellar birth rates. The association between stars and arms is transient, as we've seen, but it fits into the category of 'long-term-transient'. From the perspective of the transience of humans, the arms might as well be permanent. We will remain in the Orion arm for our lifetimes, and indeed probably for the lifetime of humanity as a whole.

So, we have the Orion constellation, which isn't real (in three dimensions), and the Orion arm, which is. But what's the relationship between the two? This is an important question because the answer to it should help us to connect observational and conceptual astronomy. But the answer is far from obvious.

To make headway on this issue, we need to have some idea of the dimensions of the Orion arm. Recall that the galactic disc is about 100,000 light years across. Some of the arms have lengths that exceed this figure, just as in a snail shell the length of the tube that coils to form a spiral as it grows exceeds the diameter of the resultant shell. But the Orion arm is much shorter, hence its alternative designation as the Orion spur. It's very roughly 10,000 light years in length and 3000–4000 light years across. We are situated about halfway along it, and slightly nearer to its inner edge than its outer in terms of breadth.

The Orion *constellation*'s closest major star to us, the top-right Bellatrix, is only 250 light years away, so it must be within the Orion arm. Suppose, however, that 'very near to' Bellatrix, from a human stargazer's perspective, there is a faint-looking star that is actually huge but very distant – say 25,000 light years away, in other words 100 times as distant as Bellatrix. Although this (fictional) star is in the Orion constellation (2D view), it cannot be within the Orion arm (3D view). A star at intermediate distance, such as the belt star Alnilam (2000 light years away) might or might not be in the Orion arm – this depends on its direction. So, in which direction are we looking when we gaze at Orion's belt? The easy answer, which is just an observational one, goes something like this. An observer who looked at Orion's belt one late evening in the middle of January from the Kielder Observatory in northern England, very close to the Scottish border, would have been looking south.

Although the observational answer is the easy one in that it can be measured by eye, it is context-dependent as the above example suggests, and changes with both time and place of observation. As the Earth spins on its axis, the direction in which we have to look to see Orion's belt changes. In the course of one night at Kielder, it changes from east to south to west, as do the directions of many other stars. However, at Kielder that night has to be in winter, because in summer Orion is nowhere to be seen from that location.

Now we look at the other side of the directional question. In what direction are we looking, in *galactic* terms, when we gaze at Orion's belt? The answer now is 'outward' in the sense that we're looking towards the Perseus arm, the next one out from our own, rather than 'inward', towards the galactic centre. This answer is *not* context-dependent: it is true of an Earth-based observer looking at Orion's belt from any latitude at any time of the night in any month of the year in which the belt is visible.

What this means is that the main stars that make up the constellation of Orion are all within the Orion arm – because we would have to look 'outward' by more than 2000 light years to look beyond this arm into 'inter-arm space'. But this correspondence between constellation and arm names cannot be guaranteed as a general rule – in fact, it is the exception. For example, the brightest stars in the constellation of Perseus are not in the Perseus arm; rather, they too are in the Orion arm. Likewise, the brightest stars in the constellation of Sagittarius are not in the Sagittarius arm, but again in the Orion arm. Indeed, the vast majority of stars that can be seen with the naked eye are located within this local arm that is our home.

There are millions of stars and planets within the Orion arm. These planets are the best hunting grounds for life, given their relative proximity to us, which is why the most successful planet-hunting device to date, the Kepler Space Telescope, was pointed into the long axis of the arm during its main exoplanet search mission, which lasted from 2009 to 2013. From the solar system 'into the long axis' does not uniquely define the direction in which Kepler was aimed, because an axis (like an axle) points in two directions. Kepler's direction of view along the Orion arm can be expressed in two ways. From a stargazing perspective, Kepler's direction of view was towards the constellation of Cygnus the swan; its field of view overlapped the constellations Cygnus, Lyra, and (to a lesser extent) Draco. From a larger-scale perspective, Kepler's direction of view was what we might call 'spirally inward'. In other words, if we were to follow the Orion arm in the direction in which Kepler was pointing, we'd end up at the centre of the Milky Way, not at its periphery. Kepler discovered more than 3000 planets within the Orion arm. Whether any of these are homes to life remains to be seen, though the answer may turn out to be 'several' of them – we'll return to this issue in later chapters.

There's a time dimension to the galaxy as well as the three spatial ones. And this fourth dimension may well affect the likelihood of life, so now we must examine it in some detail. Our portrait of the Milky Way, and of constituent parts of it like the Orion arm, is made doubly difficult to paint because of changes over time. Leonardo is not only trying to paint the Mona Lisa from within her head, but he also has the problem that she is ageing as he paints.

## Galactic Evolution and Life

The Milky Way, like other galaxies, contains stars of various ages. Its oldest stars, some of those in the globular clusters, are more than 13 billion years old, thus putting their origin only a fraction of a billion years after the origin of the universe in the Big Bang. The youngest stars in the galaxy are those that have been born only millions of years ago rather than billions. The open clusters generally contain young stars. Those of the Seven Sisters are thought to be about 100 million years in age – or 0.1 billion if you prefer, to make a comparison with stars of globular clusters more straightforward. The Orion Nebula star-forming region contains stars that are younger still. The stars making up the group called the Trapezium within this region are less than 1 million years old (0.001 billion). Others that are just switching on their nuclear furnaces about now can be regarded as having an age of zero.

Star age can affect the likelihood of life in different ways. One important feature involved here is metallicity. We need to employ this term in the sense used by astronomers rather than chemists. So, for the purposes of this book 'metallicity' means the proportion of the mass of a star that is made up of elements heavier than helium. For an astronomer, hydrogen and helium are non-metals, while all the other elements (more than a hundred of them) are metals. This connects with the difference between elements made during the phase of primordial nucleosynthesis, in the first half-hour of the universe's existence, and those that were made much later, generally during the deaths of stars. The only slight complication to this otherwise neat connection is that very small amounts of the third-heaviest element, lithium, were also made during the universe's first half-hour.

The metallicity of stars varies considerably. That of the Sun is about 0.02, which means that about 2% of its mass consists of elements heavier than helium. The range of metallicity among other stars in the Milky Way ranges from less than a thousandth of the Sun's figure to about two or three times it. Thus the Sun is at the more metallic end of the spectrum, and it's a good idea to ask why.

To answer this question, we need to examine galactic evolution as follows. Consider a star that forms a mere few million years after the Big Bang. And suppose that it is the very first star to form in its local region of the universe. Such a star will have a metallicity of zero, or extremely close to zero – the qualification being necessary because of the possible existence at that very early point in time of trace quantities of lithium. We'll assume that this star is very large, in which case its lifespan will be very short (more on this relationship later), and that it explodes as a supernova after about 10 million years.

Now consider another star born in the same region of space about 20 million years later. This region will have been enriched, in terms of metallicity, by the death of the earlier star, and any others that were born and died slightly later than it. The enrichment occurs because in the final stages of a star's life it runs out of hydrogen fuel, but, because its core temperature gets progressively hotter, it begins to be able to use helium as a fuel. It produces carbon from helium, and later still produces even heavier elements than carbon – potentially all the way up to iron. Elements heavier than iron are produced in the supernova explosion that finally tears the star apart and blasts most of its contents into the surrounding interstellar medium, from clouds of which new stars are born.

This process can be regarded as a cyclical one, continuing almost indefinitely as the galaxy ages. But it does not occur in discrete stages, and so it is not a good idea to think of 'first-generation' and 'second-generation' stars. The pattern of the birth of stars over time resembles more closely the pattern of human births (non-seasonal) than the births of, say, garden birds in the temperate regions of the Earth, which occur in a very seasonal manner, with nearly all the young birds hatching in the spring and early summer.

Although the expression 'generation I stars' is rarely found in the literature, the related expression 'population I stars' is frequently found. Personally, I tend to avoid it for the following reason. The idea behind population I and II stars, and a hypothetical population III, is as follows. Population I stars are those of the Milky Way's disc, generally younger than 10 billion years and with metallicities much like the Sun, which is one of them. Population II stars are found in the galactic halo, are older, and have lower metallicities. Population III stars, yet to be confirmed, are those earliest stars whose metallicities were close to zero.

You can see why, as a biologist, I don't like this system of naming. The first stars are given the highest population number, the youngest the lowest number. Also, if we think of the variation in metallicity, the lowest population number has the highest value. This seems perverse. So, we should acknowledge those aspects of reality that this naming system relates to, such as the typically greater age of halo stars than disc stars, but ignore the system itself.

Let's now shift our attention to the importance of metallicity for life. The life-forms that we know of have much carbon, nitrogen, oxygen, and phosphorus in their bodies. In the earliest stages of the evolution of our galaxy, these elements didn't exist in the interstellar clouds of gas from which stars formed. Somewhat later, for example 20 million years later as in the above example, they were present in very small quantities. Later still, such as when our Sun formed about 9 *billion* years after the Big Bang, they were present in much larger quantities.

The metallicity of the interstellar medium at any time will be reflected in the proto-stars that are being born at that time, and also in the proto-planetary discs that surround them. The amount of any element in the proto-star and its associated disc will in turn be reflected in the planets that form from that disc. If the disc has little or no carbon, that restriction will apply to all the planets, moons, asteroids, and comets that are generated from it. No new carbon can be synthesized in any of these bodies. And it will not be generated by the young sun until it has become a very old sun. Thus the earliest-formed planetary systems in the Milky Way, and in other galaxies too, were not fertile places for the origin of life.

It's not clear exactly when in the history of the Milky Way the first planetary systems that were potentially fertile arose. One possibility is that this point in time coincides with the difference in age between typical halo stars and typical disc stars – the 'cut-off' is in the region of about 10 billion years ago, that is, when the universe was about four billion years old. If this is true, then there may be planets on which life originated five billion years before the birth of our Sun. That's a sobering thought.

# 3 THE LIKELIHOOD OF OTHER TREES

## Defining Life Beyond the Earth

We asked the question 'what is life?' in Chapter 1. And we arrived at a tentative answer in terms of a working definition of life based on the characteristics of variation, reproduction, inheritance, metabolism, and (perhaps) being membrane-bound. Notice that this definition says nothing of chemical composition. If only the traits named above are included, there is nothing to say that, at least in theory, life should not be based on some element other than carbon.

Perhaps before going any further it's worth asking what it is that we're really looking for, in terms of life-forms scattered through the Milky Way. We're probably somewhat split in this respect. People involved in SETI – the search for extraterrestrial intelligence – have a narrower focus than astrobiologists, who would be fascinated to find anything from extraterrestrial bacteria upward. Bacteria on other planets are likely to be no more intelligent than their counterparts on Earth; nevertheless, they probably have all the defining features of life that we've been using in relation to the terrestrial biota.

Notice that intelligence is not part of the definition of life. Indeed, most life-forms on Earth do not possess it. Out of the several million species of Earth creatures, the number of species to possess intelligence ranges from 'just one' to 'a few thousand', depending on how we define it. If we use the ability to send and receive radio messages as our criterion, only our own species qualifies – and it wouldn't have done so in Darwin's day. If instead we extend the definition to 'tool use', we get to a qualifying number of a few thousand species, though even this large number represents less than 1% of the total. Tool use is associated with having a brain. It is lacking in kingdoms other than Animalia. Plants, fungi, brown algae, and all the various groups of microbes don't exhibit such behaviour. And in those animal groups in which there is no nervous system – for example sponges – or nerves but no brain – for example jellyfish – tool use is unknown.

Since there are life-forms that are not intelligent, it seems worthwhile to ask the question: are there non-living entities that *are* intelligent? The existence of the currently burgeoning field of science and technology called AI (artificial intelligence) suggests that there are. In terms of tool use, many robotic devices are now adept at some form or other of this. However, arguably no such device has yet constructed its own tool, in the way that early humans constructed tools like hand-axes from stones. And certainly no robot has, of its own volition, sent or received a radio signal. At present, such things are only possible upon human instruction, though perhaps in the future this restriction will no longer apply.

The conclusion to this line of thought is that our searches for life and for intelligence are rather different ventures. And they are conducted in rather different ways. The search for radio signals, which perhaps could come from very advanced organisms or AI systems on another planet, is quite distinct from the search for reproducing, metabolizing entities that may have no intelligence whatsoever. In SETI, we listen for intelligence-generated radio waves and we send out messages of our own. In SETL – an acronym I've not seen before, but that's no reason not to use it – we search for a certain kind of planetary atmosphere.

These two types of search carry with them different spatial limitations. If a radio source is strong enough, it can traverse the galaxy, albeit it will degrade with distance. In contrast, our ability to analyse exoplanet atmospheres is restricted to the closest such bodies – at the moment a reasonable maximum yardstick for this quest is roughly 1000 light years, which is only one-hundredth of the Milky Way's diameter (more detail on this yardstick later).

## From Life to Trees of Life

If life exists elsewhere in the Milky Way, should we expect the various forms of it that coexist on any particular exoplanet to be related to each other in the form of an evolutionary tree, like the tree of life on Earth? I'd argue that the answer to this question is 'yes', though with three qualifications.

First, it's entirely possible to have more than one tree of life on the same planetary body. In such a case they could be simultaneous, sequential, or partially overlapping in time. On Earth at present there is almost certainly only a single tree. We are pretty sure of this because DNA studies reveal a pattern of natural classification that is consistent with all organisms on our planet having originated from a single common ancestor in the distant past. This means that the existence of a 'shadow biosphere', consisting of thinly scattered microbes (referred to collectively as 'shadow life') that did not share a common ancestor with the rest of us, is hard to accept – despite its having been promoted by the physicist Paul Davies and several other scientists. We can't rule out the possibility that a few such life-forms exist in remote places on Earth and haven't been discovered yet; but in the absence of any such discoveries this hypothesis should be excised with Occam's razor. However, the fact that there is probably only a single tree of life represented by the biota of the present-day Earth does not imply that there was only ever a single origin of life on our planet. It's possible that there was an earlier origin and radiation of life, all of whose descendants were extinguished before our own tree originated – this is the 'sequential' case. It's also possible that there was an earlier origin of life, and that at least some of its descendant lineages were still in existence when our own tree originated, but that those descendant lineages later perished, either due to competition with our ancestors or for other reasons – this is the 'partially overlapping' case.

Second, it's possible – though perhaps very unlikely – that on an exoplanet somewhere in the galaxy there is a line of life rather than a tree of life. This would occur if the very first metabolizing, reproducing proto-cell gave rise to a lineage that never branched. A key feature generating branches in the tree of life on Earth is geographic separation. Often, one parental species divides into two or more daughter species – a process called speciation – because some geographic feature restricts reproductive contact between populations of the parental species that have come to inhabit different regions. In a classic book on evolutionary biology published in 1974, Harvard-based biologist Richard Lewontin said (p. 161): 'If there is any element of the theory of speciation that is likely to be generally true, it is that geographical isolation and the severe restriction of genetic exchange between populations is the first, necessary step.' Consider then a planet most of whose surface is uninhabitable but on which there is a small patch somewhere that is like a single oasis in a vast desert. If life should arise here in the form of reproducing proto-cells, just a single lineage might descend from such an ancestor. It might go forward a long way in time; it might also change in its characteristics over time; but it might never split.

Third, as well as the possibilities of dual/multiple-tree life and single-branch life, there is the possibility that the pattern of relatedness among the different life-forms on an exoplanet is more reticulate than tree-like. This would happen if horizontal gene transfer and hybridization between closely related species are even more prevalent than on Earth, to the extent that convergences between lineages rival divergences in their frequency of occurrence. In this case we might want to refer to a net of life instead of a tree of life, though in some sense the tree is still there – you cannot have convergences between branches until you have branches in the first place.

In the end, fascinating as trees of life are, and likely as it is that life on another planet will be tree-like to a degree, what we are looking for across the Milky Way, and ultimately across the universe, is life itself. We're looking for living organisms of any kind. We're especially but by no means exclusively interested in those that can be described as intelligent. And we have a suspicion that carbon-based life may be the norm, though we can't be sure. Against this background, let's begin to narrow down the search by excluding parts of the galaxy that are probably lifeless.

## Barren Parts of the Galaxy?

We'll start in the interstellar medium – that rarefied haze of gas and dust that extends across the Milky Way in the space between one star and the next. Unless life can take a form that is *very* different to what we're accustomed to, this major spatial component of the galaxy is barren indeed. But it's not homogeneous. Clouds of greater density of atoms and molecules are found dotted around here and there, surrounded by the thinner intercloud medium. And some clouds are thicker than others: dense clouds have more matter per unit volume than diffuse clouds, as their self-explanatory names tell us.

Sometimes clouds are found in complexes rather than as single puffs of celestial smoke. An example is the Orion cloud complex. When photographed with long exposure, this appears to swirl around the entire constellation. The 'appears to' is a reminder of our two-dimensional view of the sky. In fact, this cloud complex is about twice as far away as the brightest stars in Orion, Betelgeuse and Rigel. With the naked eye, or better still with binoculars or a small telescope, we can see the Orion Nebula, which is just one tiny part of the overall cloud complex – a particularly dense part that, as we noted earlier, is an active star-forming region.

Although dense clouds have many molecules, while a patch of the intercloud medium has few or none, these are relatively simple molecules such as carbon monoxide and ammonia. There are no macromolecules in these clouds. There is thus no basis for life even approximately as we know it. So in the end we rule out *all* of the interstellar medium as a home for life. And that means in spatial terms that we have ruled out more than 99% of the galaxy. (Naturally, science fiction is less constrained than science: gaseous alien life-forms feature in *The Black Cloud*, by Fred Hoyle, and in *Evolving the Alien*, by Jack Cohen and Ian Stewart.)

Next, we rule out suns. This means all suns and all parts of them. No metabolizing, reproducing life, whether simple, like bacteria, or more complex, like mammals, could exist in such a hellish environment. The temperature of the surface of our local Sun is about 6000 kelvins. Its core temperature is estimated to be about 15 million. With some variation, the situation is the same in all the other suns in the Milky Way – thousands of degrees at the surface, millions in the core. By ruling out suns as possible homes for life, we rule out more than 99% of the matter of the galaxy, having already ruled out 99% of its space.

But wait a moment. What I really mean by that last statement is that we have ruled out more than 99% of *normal* matter. Can we rule out dark matter too? This is a harder question, as we don't know what it is, and there are even some scientists who doubt its existence. The problem here is that, assuming it's real, we don't know what kind of particles it's made of. The main candidate at the time of writing is the WIMP – weakly interacting massive particle – something that's much larger than a proton or neutron, and that doesn't interact with light. Hence we can't see it. Various attempts to detect it have so far failed. Given this state of affairs, I'd say that the best conclusion regarding possible dark-matter life is: probably not, but we really don't know.

So, excluding all of the Milky Way's interstellar medium, all of its suns, and provisionally all of its dark matter too, what are we left with in terms of possible homes for life? The short answer is *planets*, but let's not restrict our focus to these just yet. The question then becomes: what else does the galaxy contain?

Here's a selection of other objects that seem likely to be barren. First, dead stars, including white dwarfs, neutron stars, and black holes. Second, those entities somewhere in between a small star and a large planet that we call brown dwarfs. These are probably failed stars in the sense that they never became hot enough to start fusing hydrogen into helium. A few brown dwarfs are known to have planets orbiting them – these too seem unlikely homes for life, given the lack of light energy produced by their quasi-suns. Third, pulsars. Although these are in fact a sub-category of neutron stars, it's worth mentioning them separately because of the famous pulsar-generated LGM signal, with this set of initials standing for Little Green Men. A pulsar is a rapidly rotating neutron star that emits a strong beam of radiation that we only see when it is pointing in our direction. It was the regularity of the beam pulses, every 1.3 seconds, that led its discoverers – Cambridge-based British astronomers Jocelyn Bell Burnell and Antony Hewish – to jokingly coin the LGM term back in the 1960s when they discovered what turned out to be the first pulsar. There are unlikely to be little green men or any other life-forms on pulsars (or other neutron stars), because these objects have a completely collapsed atomic structure.

So, after these many rulings-out of possible homes for life, what are we left with? The answer: planets, and other bodies that, like planets, orbit stars. Now it's time to zoom in on this assortment of stellar orbiters. But it will just be a preliminary zoom – one that will be expanded on as we dig deeper into planetary systems in Part III.

## Planets and Their Kin: an Overview

One of the most important lessons that we've learned from the discovery of exoplanets over the last thirty years or so is that our own planetary system is not replicated in the arrangement of planets around other stars. Our pattern of an inner four rocky planets and an outer four gas or ice giants is not a recipe that has typically been followed elsewhere, though probably in the fullness of time we'll discover this pattern around some other stars, simply because there are so many stars in the galaxy as a whole. Many of the earliest exoplanets discovered were of the type now called 'hot Jupiters' – that is, large gas giants orbiting very close to their host stars. This type of planet is absent from our own system, as is the type of 'super-Earth' that has a mass at the high end of the distribution for rocky planets – about 5–10 Earth-masses.

This finding of immense variety in the nature of planetary systems contrasts markedly with the more restricted variety in the nature of the stars that planets orbit. Stars range in mass from tiny red dwarfs to mammoth stars like Alnitak in Orion's belt. In fact, Alnitak turns out to be a multiple star system, though it looks like a single star to a naked-eye observer. Overall, the Alnitak stellar system is about 30 times as massive as the Sun, and the largest single star in it is more than 10 times the Sun's mass. If we look further afield in the Milky Way than the constellation of Orion, we find that there are also a few mega-stars with masses about 50 times that of the Sun.

Despite this variation, the main characteristics of stars are related in predictable ways. For example, a star's mass, its intrinsic luminosity, its surface temperature, and its perceived colour are all related. Massive stars are typically hotter, more luminous, and bluer than stars at the other end of the mass spectrum – red dwarfs – which are cooler, less luminous, and (naturally) redder. There are complications that set in at the end of a star's life, but the above generalization is true of all main-sequence stars, and a star spends most of its life in this phase. The term 'main sequence' is a reference to a common graphical plot of stars, one version of which has the vertical axis as intrinsic brightness (luminosity) and the horizontal axis as temperature. Such a plot (Figure 3) is called a Hertzsprung–Russell diagram, after the Danish and American astronomers who devised it.

There is no corresponding diagram for planets. Gas giant planets can orbit close in to their star, far out from it, or somewhere in between. So can small rocky ones. Rocky planets can be geologically active, like the Earth, or geologically quiescent, like Mars. Some planets have rings, others don't. Some planets in our own system, and doubtless in others, orbit with only a slight tilt of the axis compared to the perfect case of an axis at right angles to the orbital plane of the planet concerned – Jupiter is a good example, with an axial tilt of about 3 degrees. Others, such as

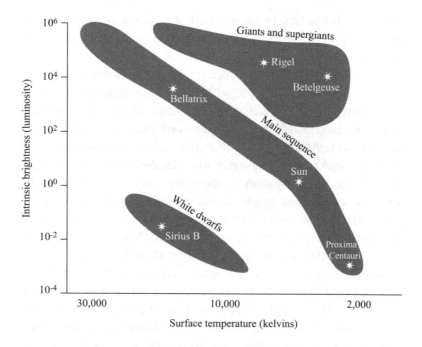

**Figure 3** The Hertzsprung–Russell diagram, showing how the intrinsic brightness (luminosity) of a star and its surface temperature are related, both for the majority of stars – the 'main sequence', shown as a downward diagonal – and for white dwarfs (below the main sequence) and giants + supergiants (above the main sequence). A star typically spends about 90% of its lifespan on the main sequence, before swelling into a giant or supergiant, and then shrinking to a white dwarf or something even smaller and denser (a neutron star or black hole). Luminosity: multiples of the Sun.

the Earth, have bigger tilts; and Uranus is an example of a planet that is effectively lying on its side.

The reason for this difference between the predictability of variation among stars and the unpredictability of variation among planets is to be found in the relative importance of historical accidents in the two cases. The mass of a sun is accidentally determined in the sense that as a giant molecular cloud collapses in on itself it usually fragments, and each fragment or sub-fragment may give rise to a star. The more massive a fragment, the more massive the star it gives rise to. But then nuclear fusion takes over, and so most of what follows is law-like rather than accidental.

In contrast, an early-stage planetary system is more complex than a snowball fight in a beer garden. There are objects of various sizes flying around, crashing into each other, breaking apart, fusing together, and moving in orbital position in both inward and outward directions. The Moon is peppered with craters partly as a result of this early snowball fight. Uranus was probably knocked over in a collision with another large body. And the Earth–Moon system may itself have arisen via collision and fusion of two proto-planets. Although our solar system now seems a model of stability, with a few exceptions like occasional out-of-control asteroids, it was once a very different sort of system, with quasi-random processes prevailing over law-like ones.

The Milky Way probably has more planets than stars – maybe more than a trillion of them. They are to be found in planetary systems that consist of one, two, three, four, five, six, seven, and eight planets. In addition to our own eight-planet system, another was discovered in 2017, called Kepler-90. And there are probably systems with nine or more planets – we just haven't found them yet. Planets orbit almost every type of star, from giants to dwarfs, including binary stars, of which there are many. In binary systems, planets can orbit one star, the other, or both together – in the last situation the planets are said to have a circumbinary orbit.

To add to the complexity and unpredictability of planetary systems, there's an assortment of objects that I've included as the 'kin' of planets in the heading of this section. They all orbit the central sun (or suns) as planets do, but they may do so directly or indirectly. The former category includes dwarf planets, asteroids, and comets. The latter includes moons, which orbit their sun indirectly by orbiting their host planet and thereby following its stellar orbit too. In our own system, the number of moons varies from none (e.g. Venus) to about 70 (Jupiter; the four large Galilean moons plus lots of smaller ones). It seems unlikely that life will be found on asteroids or comets, due to the lack of an atmosphere; but moons provide another possible type of home for life. We already noted the possibility of life in the yet-to-be-explored sub-surface oceans of Europa and Enceladus. And Saturn's largest moon Titan has an atmosphere. Its major component is the same as Earth's – nitrogen – but apart from that it's very different; for example its second-commonest gas is methane rather than oxygen.

## Rare Earth, Common Earth

The 'Rare Earth hypothesis' was proposed in the 2000 book *Rare Earth* by American scientists Peter Ward and Donald Brownlee. The nature of the hypothesis is captured by the book's subtitle: *Why Complex Life is Uncommon in the Universe*. Of course, if 'complex' is undefined, the hypothesis is very vague. Complexity is a continuous variable, not a binary one. Life does not divide into 'simple' and 'complex'. However, in their Introduction, the authors help their readers in this respect by saying what they mean by complex life – they consider it to be 'animals and higher plants'. As a biologist, I'm not sure what 'higher plants' are, but perhaps we can equate them with 'flowering plants', the group officially called the angiosperms. So the Rare Earth hypothesis that Ward and Brownlee espouse is not just that intelligent life is rare but that all animal life is rare too, as are some kinds of plant life.

Other scientists have argued that Earth-like planets and complex life may be common – see, for example, the 2018 book *One of Ten Billion Earths*, by American astrophysicist Karel Schrijver. But as with simple–complex, rare–common is a continuum. The best way out of this definitional trap is to use numbers. I'm going to do that in a very basic way now; and in a more refined way, using the Drake equation, later (Chapter 12). Let's start by assuming that the Milky Way has about a trillion planets, as suggested earlier. How many of these are potentially habitable? In our solar system the fraction is one in eight. Across all planetary systems, let's go with a pessimistic guestimate of 1 in 100 and see where we get to.

Note, however, that 'potentially habitable' and 'actually inhabited' are not the same thing. One of the great unknowns in astrobiology is the fraction of the former that become the latter. We know that on Earth the transition happened about four billion years ago, when our planet was very young. It's tempting to think that this transition is highly probable on other Earth-like planets and typically occurs at a similarly early stage. One line of argument for this high probability is the Mosaic hypothesis, as follows: every habitable patch of a planet's surface is a quasi-independent place where life may originate. Although in any one such patch the probability of an origin of life is low, taken together the probability of an origin of life occurring somewhere in the whole mosaic of patches is much higher. If there are enough such patches, the probability approaches unity. And only one place has to be successful in this respect; life, once established, may spread across the planet.

If this is true, then with a few other provisos that we'll get to later, there are probably several billion planets with life in the Milky Way. But on what fraction of them is there complex life in the sense of 'animals and higher plants'? One way to contemplate this unknown is via what we know of Earth's history. Our planet has had microbial life for more than three-quarters of its existence, animal life for less than one-quarter. When we observe current habitable exoplanets, we look at a cross-section of ages – young, middle-aged, and old. The youngest are unlikely to have animal life, but on all the others it's entirely possible.

In terms of animal life being actual rather than just possible, we can use the same argument as for the origin of microbes. The surface of the Earth – now at a later stage in its history – can be considered as a mosaic of patches in each of which animals could originate. So again the low probability for any one patch translates into a much higher probability, perhaps a near-certainty, for the Earth as a whole. Even if the probability of animals evolving from microbes on an 'average' inhabited planet is only 1 in 100, and the average duration of animal life is only a tenth of that of microbial life, we arrive at the conclusion that if there are several billion planets with microbes, there are probably several million with animals. So, complex life may in a sense be both common and rare. The number of planets with animals (or animal-equivalents) is huge, but the fraction of planets with animals is tiny. I'm going to emphasize the former rather than the latter – since the latter is uncontentious anyhow – and I'll call this view the Common Earth hypothesis. We'll expand on it in Part III.

## Parallel Trees?

When the dinosaurs became extinct about 65 million years ago, mammals underwent an adaptive radiation of forms. It's interesting to compare the nature of this radiation in Eurasia and Australia because, 65 million years ago, the starting point for it in the former landmass was a small placental mammal while in the latter it was a small marsupial. There was no land-bridge between the two landmasses then, though Australia was joined to Antarctica.

Mammals in Eurasia diversified into various forms, including large carnivores such as wolves, underground burrowing forms such as moles, gliding forms such as colugos, and flying forms – bats. Interestingly, although the radiation of marsupial mammals in Australia was an entirely independent process, similar forms often resulted from it. These include marsupial wolves, marsupial moles, and marsupial gliders, though no marsupial bats – the reason for this 'omission' is unclear.

This similarity in the range of forms produced by the independent radiations of two sub-trees (or large branches) of the tree of life on Earth is referred to as parallel evolution. Its existence immediately raises the question: can parallel evolution also occur between the overall trees of life on different planets, producing similar arrays of creatures? Such trees must be completely independent of each other, assuming that our rejection of the Panspermia hypothesis in Chapter 1 was justified.

'Similar' in this context is just as problematic as 'common' and 'complex'. Like them, it's a reference to an undefined position on a continuum, this time from identical to very different.

So, how similar is similar? The limiting case in one direction is an identical array of species on 'exoplanet X' to those found somewhere else, for example on Earth. The probability of this seems vanishingly small. The limiting case in the other direction is trees diverging right at the start. An example approaching this latter extreme would be a planet on which, although large organisms evolved, none of them had bodies divided up into cells. Such organisms exist on Earth – the acellular slime moulds – but they only represent a small branch of our tree of life, about 1000 species out of several million.

It seems likely that when we first discover a tree of life on another planet it will fall somewhere between these two extremes. However, if there really are trees of life on millions or billions of planets, as proposed in the previous section, we should expect many different degrees of similarity in inter-tree comparisons.

One factor that has an influence on the degree of similarity between two branches of the tree of life on Earth, such as the parallels between the marsupial and placental mammal radiations, is the range of habitats available. If two regions of the Earth both have soil, a ground surface, and trees, then it's not surprising that in both places burrowing, running, and gliding forms should evolve. But habitats aren't inert things that organisms adapt themselves to – a maple tree is both a life-form in its own right and a habitat for many others. Even a mammal can be a habitat – for example for a flea or tapeworm. So some parts of habitats evolve along with their users. And the whole process takes place in what we call the biosphere – the 'thin sliver of existence' to which we now turn.

*Part II*

# Life Here, Implications for Elsewhere

## Key Hypotheses

### The Parallel Places Hypothesis
The ubiquity of gravity and topography, coupled with the presence of surface water, means that the broad range of habitat types on most inhabited worlds will parallel that of the Earth.

### The Ubiquitous Skeletons Hypothesis
Large organisms, whichever planet they are found on, possess skeletal structures, whether bone, wood, or 'other', which enable them to resist the downward force of gravity.

### The Unnecessary Intelligence Hypothesis
The way in which evolution works means that there is no guarantee that intelligence will evolve on any particular planet. The relationship between intelligence and fitness is complex, as shown on Earth by evolutionary trajectories that lead to the loss of a brain.

# 4 A THIN SLIVER OF EXISTENCE

## Deep within the Earth

Although we think of the Earth as an inhabited planet, this is, literally, a superficial view. The roots of life extend much deeper in time than they do in space. Taking a three-dimensional approach, the vast majority of our planet is devoid of life-forms. To appreciate the truth of this statement, it's useful if we compare distances across the planet with those down into its interior. For example, the distance from New York to Berlin is almost exactly the same as the distance from New York to the centre of the Earth. On the former journey, there are life-forms all the way. Most are small, such as those of the plankton, an invisible mass of floating life that extends across the Atlantic. Others are large, like the trees of a European forest. And many are humans, especially on land. Organisms would be encountered every step of the way by hypothetical travellers using bicycles and canoes, with microscopes in their backpacks.

Now contrast this scenario with its equivalent on the long journey down towards the centre of the Earth. Naturally, our travellers on this journey are even more hypothetical than those on the first. This time we should perhaps imagine a kind of small submersible vehicle with a drill on its underside so that it can go down through solids as well as liquids. But again its occupants have microscopes so they can see bacteria and other tiny forms of life.

Suppose the starting point is in Long Island Sound, just offshore from Stony Brook. The first part of the trip is easy, though soon the drill needs to be employed and the going gets tough. However, our job here is not to consider the difficulties of travelling but rather the analysis of a series of samples of material taken from outside the vehicle. In general terms, the results are known, even though the journey, as I've described it, has never been undertaken. On the way down, the abundance of life would change from lots to some to none. But when would it reach a clear 'none'? Although we can't give a precise numerical answer to that question, we can give an estimated one in proportional terms. For more than 99% of the journey, in other words for all of it except its very earliest stages, there would be no life at all.

One reason for this is temperature. This increases going down, and it does so rather rapidly. Moving from fictional journeys to actual drilling experiments, the temperature at 10 kilometres (6 miles) down reaches about 250 degrees Celsius. Given that none of the heat-loving microbes that we call hyperthermophiles can tolerate temperatures above 150 degrees, we can be fairly confident that there is no life at all at this sort of depth. If the deepest microbes are about 6.4 kilometres (4 miles) down, and they probably aren't as deep as that, then, since the distance to the centre of the Earth is roughly 6400 kilometres (4000 miles), microbes extend down for only 0.1% of the Earth's radius. So our ballpark estimate of 99% of the journey being barren can be revised to 99.9%.

A few drilling projects have reached depths of more than 10 kilometres. The deepest to date, the Kola ultra-deep borehole in northwest Russia, penetrated more than 12 kilometres (7.5 miles) below ground. While the conditions for life simply don't exist so far below a continental landmass, they certainly do exist in some parts of the world a similar distance below the surface of the sea. The deepest of all the ocean trenches on Earth is the Mariana Trench in the Pacific Ocean (northwest of the island of Guam), which reaches a maximum depth of almost 11 kilometres (6.8 miles). Various life-forms are found in this and other ocean trenches, and doubtless within the underlying substratum. So in the end our estimate of the maximum depth of life should be somewhere between 0.1% and 1% of the Earth's radius, leaving the remaining 99% to 99.9% barren.

This exercise has taken us to the lower limit of the biosphere – the thin envelope around our planet in which all life-forms exist. In the next section we'll look at its top. But before we leave the depths of the Earth there's one further thing we should do.

Imagine again that journey to the centre of the Earth. Now suppose that on our way back up we stop halfway to the surface. At that depth we're within the region called the outer core. This, like the inner core, is made up largely of iron and nickel. But there's an important difference: the outer core is liquid. Although it's no place for life, it nevertheless has major effects on the Earth that extend right up into the atmosphere. Movement of the liquid outer core is responsible for the Earth's magnetic field. And that field is what deflects the solar wind away from most of the surface of our planet, funnelling it instead into two arrival areas around the poles, thus giving the northern and southern lights – the Aurora Borealis and Aurora Australis. Some planets have different sorts of core and hence no magnetic fields. This difference may have important consequences for life.

## High up in the Sky

Where is the top of the biosphere – in other words how high up are the highest organisms of all? This question is tricky enough with large life-forms such as birds, trickier still with small ones such as bacteria. To answer it, let's undertake another journey, this time up rather than down. Instead of a submersible-with-drill, we'll use a new version of the Space Shuttle specially designed for retrieving atmospheric samples from a variety of altitudes, all the way from the Earth's surface to the beginning of space. As we go upward, the abundance of life outside the craft will reduce from lots to some to none, just as it did on our downward journey to the centre of the Earth. Of course, its exact pattern of decline will be different, but we ask the same question as before: when does it reach zero?

The end of this upward journey is hard to define, given that our atmosphere thins out and becomes space in a very gradual way. So we have to make a quasi-arbitrary choice. Let's choose a height of 640 kilometres (400 miles), which is a tenth of the distance to the centre of the Earth. In surface terms, we're no longer travelling from New York to Berlin, but only from New York to Cleveland, Ohio. Having travelled to this height we find ourselves in the lower reaches of the outermost part of Earth's atmosphere – the exosphere. Hydrogen, rather than nitrogen, is the commonest gas here. And all gases present are so rarefied that they're barely present at all. Indeed, it's a philosophical question whether this is the last part of the atmosphere or the closest part of space. The International Space Station is below us, as is the Hubble Space Telescope.

There can be no living creatures at this altitude – at least not *actively* living ones. The presence of floating spores in a dormant condition cannot be ruled out, but seems very unlikely. So, we conclude that we are above the biosphere, and we begin our descent. How far down do we have to go before we find life?

The ozone layer is a good marker in this respect. Any living creature above it is bombarded by the full force of the Sun's ultraviolet radiation, while those below it are protected from this – and in particular from the harmful, shorter-wavelength ultraviolet. This layer, which is of variable thickness, is centred on an altitude of about 25 kilometres (15 miles). That's only about one-thirtieth of the height to which we initially ascended. The term 'ozone layer' is perhaps misleading – the concentration of ozone here is less than 10 parts per million. Nevertheless, its protective effects for life are significant (there will be further discussion of this in Chapter 15).

Is there any life immediately below the ozone layer? The answer to this is a resounding 'yes'. The band of altitudes we're talking about here is between about 10 and 20 kilometres (6–12 miles). There are humans here, albeit only in planes. There are also some birds. The Rüppell's vulture is known to be able to fly at a height of 11 kilometres – there is solid evidence for this in the form of a collision between one such vulture and a commercial aircraft over the Ivory Coast in the 1970s. There were enough feathers left in the engine to allow species-level identification.

It's known that there are airborne bacteria at this altitude – indeed even a bit higher. So, one way to define the top of the biosphere is as being roughly coincident with the ozone layer. Of course, reality is never so neat. The biosphere doesn't have a clearly defined top or bottom. Its shape is complex. It's not a sphere at all, or even an approximation to one, as we will now see.

## The Shape of the Biosphere

The use of 'sphere' for the realm of life is just a convenient shorthand notation. This is true also in relation to such usage for layers above it, such as the stratosphere (containing the ozone layer), and those below it, such as the asthenosphere (part of the Earth's mantle). The correct term for the shape that represents the difference between two concentric spheres with different radii is a 'spherical shell'. All of the Earth's various 'spheres', both solid and gaseous, are approximations to this shape, not to that of a sphere, despite their names. Ironically, the only part of the Earth that does individually approximate to a sphere is the core, whose name is sphere-free.

Regarding the nature of the biosphere's approximation to a spherical shell, we've seen some of the irregularities already, including the existence of deep ocean trenches. And high mountains such as the Himalayas are well known. Less well known are subterranean water-bodies such as Lake Vostok, in Antarctica. The surface of this lake is about 500 metres (1600 feet) below sea level. Water from the lake obtained via drill holes has been analysed and found to contain a variety of organisms, mostly bacteria. However, it is unclear whether these are genuine residents of the lake, contaminants from drilling, or a mixture of the two. Further work is planned to try to distinguish between these possibilities.

The most interesting aspect of the biosphere's shape, though, is not something that can be shown in terms of a diagram of its outermost edges. Rather, it's what we might call 'density variation' in its interior. Imagine standing in a tropical rainforest and counting organisms all the way up to the tops of the highest trees and also right down into a deep hole that was dug for this purpose. Most life occurs between a few metres down and a few metres above the treetops. The overall vertical distance between these two levels is less than that of a tall building. This is a tiny fraction of the biosphere's maximum thickness of perhaps 30 kilometres (18 miles).

What this tells us is that the thin sliver of existence around the Earth for the vast majority of organisms is far thinner than the spherical shell whose entirety we call the biosphere. On land, there is a profusion of plant life extending upward to a *tiny* degree – even in the case of the very tallest trees – from the surface of the soil. Animal life is tightly clustered in and around the surfaces of both the soil and the vegetation. And microbial life is clustered there too. In the oceans the situation is both similar and different. There is again tight clustering, but with two layers of highest density rather than one: close to the sea surface and in/on the seabed.

In deep oceans there is too little light for plant life to be clustered at the bottom, but animals densely populate the seabed nevertheless. They are divided into the infauna (buried) and the epifauna (moving over the substratum). And there are bottom-dwelling microbes aplenty. Also, plant life, animal life, and microbial life are all clustered near the surface of the sea in the form of plankton – a collective term for a wide range of types of organisms, including algae, bacteria, archaea, and various animal forms, including small crustaceans and fish larvae. In between the clusters near these two interfaces (substratum/deep sea and shallow sea/air) is a variable depth of water column with plenty of life – just less of it than above and below.

Particularly extreme density variation, both vertical and horizontal, occurs in deserts. Considering the vertical profile of life above and below a typical patch of the Sahara, for example, it extends much less far above and below ground than in a tropical forest. And horizontally there are pronounced density steps at the edges of oases. Overall, several thousand species of plants and animals can be found in the Sahara, but in all cases with an extremely patchy distribution.

Patchiness, indeed, is the norm for most species, wherever they are found within the biosphere. And if organisms are patchily distributed then so too are the ecological interactions between them. A useful closing mental image for this section is that of ecological interactions pictured as a tangle of red lines – the colour choice deriving from 'nature red in tooth and claw' (from the 1849 poem *In Memoriam A.H.H.* by Alfred Lord Tennyson). These tangles are thickest around land/water/air interfaces, and thickest too where there is plenty of humidity. They thin out going upwards, downwards, or laterally into areas of drought. And the patchiness occurs at different levels of spatial scale. In a moist tropical region the density of interactions is high in general. But picture the dramatic rise in interactions among organisms going into a termite's nest.

Ecological interactions typically involve movement; but organisms are not the only things that move. Within the biosphere there are air currents and water currents. These carry solid matter with them – for example wind-blown sand and sea-rounded pebbles. Avalanches could be considered 'land currents'. And the base of the biosphere is itself always on the move, albeit very slowly by comparison with movements within it. This is what we'll look at next.

## Plates and Crust

We take it for granted that the Moon is peppered with craters – they are easy to see with a telescope or even, in the case of the largest ones, with a pair of binoculars. But we rarely stop to ask why the same is not true of the Earth. Both bodies were subjected to heavy bombardment by large impactors during an early phase of the solar system's history; and, when this phase ended, both doubtless bore all the resultant scars. Those scars are among the craters we see on the Moon today, but we see few such scars on our home planet.

This does not mean that the Earth is crater-free. One of the best-preserved ones that can be seen today is Meteor Crater (also called Barringer Crater) in Arizona. This is 1.2 kilometres (three-quarters of a mile) in diameter and reaches a maximum depth of about 200 metres (650 feet). It was caused by the impact of a large meteorite approximately 50,000 years ago. Another well-preserved crater is located near the town of Lonar in western India. Unlike Meteor Crater, this is water-filled and generally referred to as Lonar Crater Lake. However, it is similar in size and depth to Meteor Crater, and shares with it a relatively young age.

Youth is the key to finding well-preserved craters on Earth. Well-preserved old ones, for example those whose age could be measured in billions of years rather than thousands or millions, are non-existent; though there are a few poorly preserved old ones, such as Vredefort in South Africa, which is the Earth's largest and second-oldest crater. The lack of well-preserved old craters is partly due to erosion and partly due to the fact that our planet is continually being resurfaced by the movement of large plates of solid material, the outer surfaces of which constitute the familiar continents and the (less familiar) ocean floors. The largest plates, named after the main landmasses or oceans that they carry, are the Pacific, North American, Eurasian, African, Antarctic, Australian, and South American plates. And there are lots of smaller ones too.

The theory of plate movements (or plate tectonics) was a major advance in twentieth-century geology. The broad form of this theory was established, and generally accepted, by the early 1970s, though much has been added since then, and its refinement continues. Even in the early days of the theory, it was apparent that it provided an explanation for the phenomenon of continental drift, something that had been suggested – though without a knowledge of the underlying mechanism – by various earlier authors, notably the German scientist Alfred Wegener in 1912. According to the tectonics theory, the Earth's plates move slowly (a few millimetres to a few centimetres per year), driven by convection currents in the underlying mantle. New plate material is made at creative boundaries, such as the mid-Atlantic Ridge; old material disappears at destructive boundaries, such as where the Nazca oceanic plate is being subducted under the South American plate, a process responsible for building the Andes mountain range.

The plates vary in thickness from about 10 kilometres (6 miles) to more than 10 times that figure, with continental areas generally being thicker than oceanic ones. The base of the plates does not correspond to the base of the Earth's crust. The distinction between plates and under-plate material is made on the basis of movement and associated physical characteristics such as changes in viscosity. In contrast, the distinction between crust and underlying mantle is made on the basis of the chemistry of the rocks. What moves in the form of a plate is a large section of Earth's crust plus the layer of mantle immediately below it.

So, in terms of lower boundaries to various spherical shells on the Earth beneath the atmosphere, the base of the biosphere is underlain by the base of the crust, which in turn is underlain by the base of the plates, which corresponds to the base of the lithosphere (comprising crust and uppermost mantle) – see Figure 4. We need to keep in mind that many of the boundaries between the various spherical shells that make up the biosphere are irregular rather than smooth. Recall that the Mariana Trench extends down to about 11 kilometres (6.5 miles). Here, the base of the biosphere is much closer to the base of the crust than in most other parts of the Earth – especially in continental ones where the crust is thickest.

## Forms of Division

We've spent a good while delimiting the biosphere. Now we need to delimit parts of it from other parts. This can be done in various ways, each of which is useful in particular contexts. The most obvious distinction is between marine and emergent-land components, with the former representing about 70% of the Earth's surface, the latter 30%. And from the point of view of organisms that live there, we should divide the latter component into fresh water and dry land. (The land part is often called 'terrestrial' in ecology, but I'm trying to avoid dual usage of that term, given the astrobiological scope of this book.)

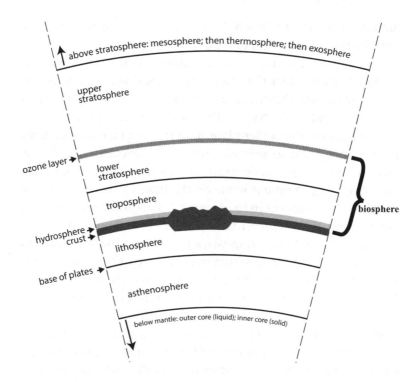

**Figure 4** Cross-section of the biosphere, showing its gaseous, liquid, and solid components. The vertical extent of the biosphere is from the ozone layer down to a point within the crust (and hence within the top of the lithosphere) that underlies the oceans, soil (most land areas), sand (deserts), and ice (Antarctica). Note: top-to-bottom depths of the various shells are not shown to scale, but are approximately as follows, starting from the lowest one. Asthenosphere 600 km; lithosphere 100 km (including crust 5–50 km); troposphere 10 km; stratosphere 40 km (including ozone layer 10 km). The depth of the biosphere varies from place to place but is typically between about 20 and 30 km.

The non-marine part of the biosphere can be divided into biogeographical realms – areas within each of which organisms have been coevolving with each other for millions of years, but between which they have not, owing to geographical isolation by barriers consisting of oceans, mountain ranges, or deserts. These are sometimes called Wallace's realms – in honour of the nineteenth-century naturalist Alfred Russel Wallace, co-proposer, with Charles Darwin, of the theory of natural selection. In some cases they correspond with continents (e.g. North American or Nearctic realm, South American or Neotropical realm). In other cases, realms and continents don't coincide exactly (e.g. Palaearctic realm, which includes most of Eurasia and the northern part of Africa down to the Sahara).

Realms can themselves be subdivided into biomes on the basis of climate and the growth-form of the dominant vegetation that prevailed before human alteration of the landscape. For example, most of the British Isles belong to the temperate deciduous forest biome, even though most of the land surface is now agricultural rather than forested. In Canada, some of the very southernmost strip is also deciduous forest; at higher latitude there is boreal forest, also known as taiga, where conifers predominate. And further north again, where the cold climate and frozen soil do not permit tree growth, there is the biome called tundra. Here, ground-level vegetation consists of various cold-hardy plants, including those of the heather family (Ericaceae).

There is no real equivalent, in the marine part of the biosphere, of realms and biomes. Here the emphasis is on continuity of the ecosystem rather than division. The entire 70% of the Earth that is covered by seawater is sometimes referred to as the World Ocean to emphasize this point. The shape of this ocean has changed over time due to plate tectonics and associated continental drift. We'll look at this long-term change in the distribution of land and sea in the next section.

The biomes of the land are not homogeneous: within each there are often many different habitats in addition to the one after which the biome may be named, such as deciduous forest. In the British Isles, there are many coastal habitats with unique floral and faunal characteristics. These include sand-dune habitats, where marram grass predominates, and saltmarsh habitats, where marsh grasses and other salt-tolerant plants prevail. And before human habitation some high-altitude regions were probably characterized by coniferous forests.

Although the marine environment is usually not divided into units like realms and biomes, it most certainly can be divided up

into habitats – usually on the basis of depth. We've already seen
the tendency for life to cluster around interfaces between solid,
liquid, and gaseous phases of matter. So the plankton can be
thought of as living in, and indeed constituting part of, the
superficial water habitat. The organisms of the bottom-dwelling
benthos, clustered on and in the seabed, are very different; they
include attached forms such as sea lilies, which are not a viable
option near the surface – except where the ocean reaches its
edges, and its bottom and top approach each other. Hardest of all
to classify perhaps is the intertidal habitat. The animals found
there are derived from emersion-tolerant marine forms (e.g.
barnacles) and immersion-tolerant land-based ones (e.g. the sea
slater, a type of coastal woodlouse).

Within habitats are microhabitats. Animals that live in the
interstices of the soil would not survive in the canopy of a forest
habitat, and vice versa. Equally, attached organisms that survive
in the lowermost zone of the intertidal would not survive in its
uppermost zone, and again vice versa. In freshwater habitats,
the situation is much the same. In a deep lake, the difference
between plankton and benthos parallels that of the ocean; and in
a small overgrown pond there are microhabitats that are almost
like land, such as the surface of matted broadleaved pondweeds.
For parasites, the microhabitat is often the host animal or part of
it. In the insect world, there are parasitoid wasps that parasitize
other insects, and hyperparasitic insects that parasitize the para-
sitoids. For such insects, and many other animals, the microhabitats of the adults and the young of the same species are quite
distinct. We'll examine habitats and microhabitats further in
Chapter 6.

## Going Back in Time

The biosphere that we've looked at so far is that of the present
day. Clearly, biospheres of the past were different from ours by
various degrees. Let's now try to get some idea of the nature of
this variation, by taking a series of steps back in time. The first
steps will be relatively small, the following ones giant leaps.

Ten million years ago, the main difference in the biosphere compared to today was the lack of humans. At that point in the past the lineage that would lead to both chimps and humans had not yet split. Over most of the tree of life, such within-family details don't matter much to global ecology. But they matter within the Hominidae because one species in this family – us – eventually altered the biosphere drastically. Ten million years ago, the biomes really were as their names suggested – deciduous forest was just that, rather than a swathe of agricultural land punctuated by cities and tiny bits of forest that to varying extents resemble the primeval one. But in other respects the biosphere was much the same. For example, the positions of the continents were not much different from those they occupy today.

One hundred million years ago, the dinosaurs rather than the mammals were the dominant land vertebrates. Sauropods, including the largest and heaviest land animals ever, were flourishing. Flowering plants – today's main land plants – had diversified widely, though they were less dominant than they are now. The continents were grouped into two very large landmasses – the northern Laurasia, consisting of North America and most of Eurasia, and the southern Gondwanaland, consisting of everything else.

Three hundred million years ago, there were no dinosaurs; at this stage in evolution they were just one of many future possibilities. And there were no flowering plants either – quite a contrast to today's 300,000 species of them. A biosphere without flowers is hard to imagine, but such a biosphere prevailed for more than 90% of Earth's history to date. The coal that we use from this period – the Carboniferous (coal-bearing) period – comes not from the decay of oaks and ashes but rather from tree-ferns and club mosses, among other non-flowering plants. The continents were all grouped together in a single supercontinent called, appropriately, Pangea (all-Earth).

Five hundred million years ago, there were no land animals – neither vertebrates nor invertebrates. And there were no trees either, whether ferns, conifers, or broadleaves. The world's flora and fauna were largely marine. As far as we know, the land biota was almost exclusively microbial; multicellular forms, if there were any, would have been algal and possibly fungal. As we've already seen, some algal groups were the ancestors of the land plants (red and green algae), while others represent a completely independent experiment in multicellularity (brown algae). The algae present 500 mya (millions of years ago) were red and green; the browns originated much more recently.

One billion years ago, marine red algae may have been the sole multicellular forms in an overwhelmingly microbial world. That world was characterized by a supercontinent – but not Pangea. We now recognize a long-term cycling of continents into and out of supercontinents. The name given to the billion-year-old Pangea-equivalent is Rodinia.

Two billion years ago, there were probably no multicellular life-forms at all. And the microbes (in the broad sense of single cells or loose aggregations of them) may have belonged solely to the domains Bacteria and Archaea. The domain into which all animal, plant, and fungal life fits – the Eukarya – is thought to have originated, in the form of complex individual cells, at least 1.5 bya (billions of years ago); but there is a wide margin of error around this estimate, and the true figure *could* be a little over 2 bya.

Three billion years ago, there were almost certainly no eukaryotes – though we always need to remember the old saying 'absence of evidence is not evidence of absence'. The microbial forms present were bacterial and archaeal. However, it is useful to remember that 'microbial' does not necessarily imply 'unicellular'. One key group of microbes present from before 3 bya until the present day is the Cyanobacteria. These can exist in the form of (two-dimensional) filaments and (three-dimensional) mats. Indeed it is stacks of cyanobacterial mats that form the structures called stromatolites that are among the earliest-known fossils. The reason the Cyanobacteria are 'key' is that they are thought to have been the first organisms that photosynthesized and produced oxygen on a large scale. It was through their photosynthesis that the Great Oxygenation Event occurred (about 2.5 bya: see Chapter 5). Before then, the Earth's atmosphere contained very little of this critically important gas.

Four billion years ago, the very first life-forms had either recently arisen or were soon to arise. The span of estimates for the origin of life straddles this period – with proposed figures ranging from about 4.3 to 3.7 bya. There are many difficulties in being certain that claimed traces of life from this early period really are biogenic – as emphasized in a 2019 review article by the palaeontologist Emmanuelle Javaux. So it is not clear whether there was a biosphere at all 4 bya, or merely a spherical shell around the Earth that would later become one.

We should appreciate that there are many ifs and buts relating to the above brief account of the biosphere's history. In taking large leaps back in time, we are liable to miss events that are crucial but short-lived (compared to our leaps), notably mass extinctions and ice ages. If one of our leaps back had been to about 250 mya we would have found ourselves in the middle of the biggest mass extinction event of all to affect the animal kingdom. And if another leap back had taken us to about 700 mya we'd have been in the Cryogenian period, whose name speaks for itself.

It will be important to recall this brief biospheric history when we come to consider the possibility of finding biospheres on other planets. The several thousand exoplanets discovered so far vary in age as well as in size, orbit, and other features. Of those that turn out to be potentially habitable, some will be old, others young. It would be reasonable to expect some of the old ones to have biogenic oxygen-rich atmospheres, but unreasonable to expect the same for any of the very young ones. Having said that, we are still entirely ignorant of the extent to which other biospheres will turn out to parallel our own. Perhaps the only thing that can be said for certain is that life-forms of any kind need energy – and that's where we're heading in the next chapter.

# 5 ENERGY AND LIFE

## To Eat or Not to Eat

The Venus flytrap is an interesting exception to a general rule. On Earth, the two great kingdoms that include very large multicellular creatures – animals and plants – are generally distinguished by eating and not eating, respectively. For our purposes here we define 'eating' as obtaining energy from consuming other organisms, or parts of them. But flytraps cross the divide. They self-feed, using photosynthesis, and also eat insects and other small invertebrates such as spiders. So, they are both autotrophic (self-feeders) and heterotrophic (feeding on others). And they are not alone in this respect. Although there is only one species of Venus flytrap, there are hundreds of species of insect-eating plants, some closely related to the flytrap (sundews), others not (pitcher plants). The last of these, in which insects are trapped in a fluid-filled cavity in the plant, have evolved multiple times.

Although less well known, there are parasitic plants (another form of heterotrophy). Like pitcher plants, which can be thought of as predators – albeit immobile ones – parasitic plants have originated many times during the evolution of the plant kingdom. The host for these plants may be another plant or a fungus. Some parasitic plants retain the ability to photosynthesize (and hence are combined auto-hetero-trophs, like flytraps, sundews, and pitchers), while others have lost it altogether. This latter type of parasitic plant is thus totally heterotrophic.

So, plants can be heterotrophic, and it turns out that animals can be autotrophic too. There are several groups of animals that can be called sea slugs – marine gastropods that have lost their shells. Some of these can photosynthesize as well as consume food organisms. Indeed, consumption and photosynthesis are closely related in the following way – based on one of the most-studied species of sea slug in this respect, *Elysia chlorotica*. These animals eat algae, in particular the yellow-green alga *Vaucheria litorea*, which are partially digested. However, the digestion stops short of destroying the algal chloroplasts, which are then engulfed by many of the sea slug's cells, where they continue to function as photosynthetic organelles, just as they did in their original algal home. Their lifespan in the mollusc is doubtless shorter than it would have been had they not been eaten. But this isn't a problem – dying chloroplasts in the sea slug are replaced by those eaten later, and so a kind of dynamic equilibrium is established.

For those who like neat categories with easy definitions, the above examples of heterotrophic plants and partially auto-trophic animals are annoying. We can no longer have clean definitions of these two broad categories of life-forms based on eating and not eating. However, that's not a problem, because these days we define plants as all the descendants of a particular lineage of proto-algae in the distant past; and likewise we define animals as all the descendants of a particular lineage of proto-sponges. To those who are more interested in the possibilities inherent in the evolutionary process than in erecting categories for its products, these classification-blurring examples are fas-cinating. They also point to the possibility of even more fascin-ating combinations of plant and animal characteristics in the life-forms of other biospheres. John Wyndham's large mobile predatory plants, the triffids, may still be the stuff of science fiction, but one day we may discover their real equivalents.

However, exceptions must also be seen in a balanced way. Although we cannot yet imagine the rules and exceptions that apply to other biospheres, on Earth they are clear. Autotrophy and heterotrophy are the rules; flytraps, green sea slugs *et al.* are the exceptions. On a recent forest walk, I came over a rise and stopped, fascinated by the unexpected presence of three roe deer on the track just a short distance ahead of me. They stopped too, surprised by my presence, before bounding off into the depths of the forest. In the frozen moment of our astonished stasis, there were birds overhead, grasses lining the edge of the track, and a lone butterfly perched on a dandelion flower. All these creatures were either photosynthetic autotrophs or organism-eating heterotrophs. If there were any exceptions at all to that rule in this particular ecosystem, they were hidden from my view.

It's easy to understand the business of eating, at least as manifested in animals like deer. If you hide yourself better than I did on my walk, you can watch it happen. Eating is a behavioural thing. But photosynthesizing is not. The plants doing it are not to be seen snapping up carbon dioxide molecules from the air. Rather, photosynthesis is a biochemical (and biophysical) thing. It's one whose existence we know about from reading books, though we could infer the existence of some process of this kind from observations of plants dying when kept in the dark. Since photosynthesis is so crucial, and yet from a behavioural perspective so cryptic, we need to examine it to see how it works. This will also lead in to our subsequent look at how it's possible for an organism to be autotrophic without being photosynthetic.

## Inside a Maple Leaf

Most photosynthesis in a present-day forest ecosystem is carried out in the leaves of trees. Let's pick a tree, then one of its leaves, and probe inside this leaf to see what's happening. We'll choose a maple tree. However, since 'maple' covers more than a hundred species, let's be more specific and choose the tree that's called a sycamore in the British Isles and a sycamore maple in North America, namely *Acer pseudoplatanus*. If we chose instead an oak tree or a rowan tree, the details of the photosynthetic process in the leaf would be much the same. Having chosen the sycamore, let's now examine one particular leaf at one particular time of year – a leaf near the top of the tree in summer. Choosing a leaf near the bottom wouldn't change things much;

however, changing the season to winter would – as sycamore is deciduous.

Our chosen leaf, like its compatriots, is broad and thin. This is an adaptation to maximize its light-intercepting surface area. Most plants share this leaf adaptation, though some do not – for example succulents have leaves that are fat rather than flat because they live in dry areas where the need to conserve water alters the balance of selective pressures acting on leaves as they evolve.

Sycamore leaves, and other similar broad/thin leaves, are only a few cells thick. Sandwiched between upper and lower epidermal cells are cells of the mesophyll layer (literally, middle-of-leaf). These are packed with chloroplasts, the intracellular organelles in which photosynthesis takes place. Mesophyll cells near the upper epidermis have most chloroplasts – after all, except on stormy days, these are exposed to more sunlight than the ones below them. Like many organelles within any eukaryote cell, chloroplasts are membrane-bound. Inside, there is a fluid filling and some folded compartments called thylakoids with their own membranes (Figure 5). It is within these membranes that photosynthesis takes place.

The photosynthetic process is generally written as follows: $CO_2 + H_2O \rightarrow$ glucose + $O_2$. The carbon dioxide comes from the atmosphere; the water comes from the soil via the roots and the tree's internal tubular transport system; the arrow implies the use of light energy from the Sun; glucose is an energy-storage molecule that's used to power metabolism; and the oxygen is released into the atmosphere. As we saw previously, oxygen released from photosynthesis in ancient Cyanobacteria drove up the amount of this gas in our atmosphere early in biospheric history – between two and three billion years ago.

The above brief formula is of course just shorthand notation for a very complex process. Let's now look at it in a bit more detail. If you don't want this detail, skip the next two paragraphs.

The molecule that captures the all-important light is chlorophyll. This molecule absorbs most photons from the blue and red areas of the visible spectrum, and far fewer from the

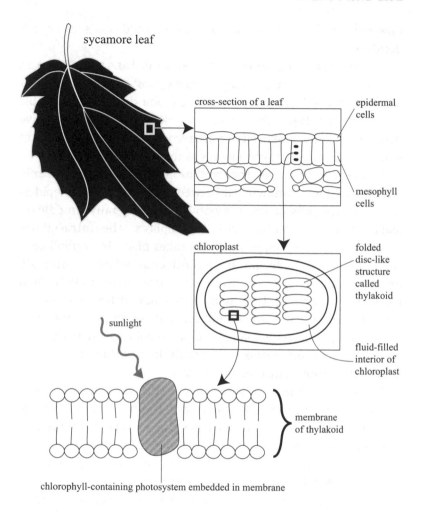

**Figure 5** Internal structure of a leaf, showing the subcellular organelles – chloroplasts – in which photosynthesis takes place. The crucial light-absorbing chlorophyll molecules are located within photosystems that are embedded in the chloroplast's internal membranes. Photosynthesis starts here, and continues in the fluid-filled chloroplast interior, where glucose is synthesized.

intermediate green. These it mostly reflects, which is why leaves containing chlorophyll appear green in colour. The chlorophyll molecules don't exist in isolation – they are part of complexes of many molecules, including proteins, which are collectively called photosystems. In these systems, which are located in the internal membranes of the chloroplasts, light is captured and water is split into hydrogen and oxygen. Energy is stored in biochemical form in the usual storage molecules (such as ATP, or adenosine triphosphate) and moved into the fluid part of the chloroplast's interior, where the inorganic carbon of $CO_2$ is turned into the organic carbon of glucose. Glucose is one of the simplest sugars – one with a single ring structure, as opposed to the double-ring that is the basis for the sugar sucrose. In chemical terms, glucose is $C_6H_{12}O_6$.

Although glucose can be used directly to power the leaf's metabolism, much of it is used elsewhere in the tree. Roots, for example, need to metabolize, and they get their glucose from the leaves. So, either glucose or a derivative of it needs to be moved around and often stored for various periods of time – after all, deciduous trees don't stop metabolizing in the winter. Storage of glucose in plants is usually in the form of starch – a macromolecule made up of lots of glucose subunits. Starch can store surplus energy for a period of time and then release it, by being broken down into glucose again.

In a forest, the glucose produced by photosynthesis in the leaves of trees and other plants powers the entire ecosystem. The food web starts with this and ends when detritivores and decomposers make use of dead plant and animal material, thus liberating nutrients back into the soil, where they are then taken up by roots. So, the nutrients really do go round in a cycle, while the energy does not. Rather, at every step in the food web, energy is dissipated as heat. Hence ecosystems can only be sustained by continued input of abiotic energy. In a forest this is sunlight, but in some other types of system it is chemical energy instead, as we'll see shortly, after first looking at a different kind of photosynthesis.

## Bacteria that Make Sulphur

As we saw in the previous section, light, carbon dioxide, and water are the three main inputs into photosynthesis in a maple tree, and indeed in plants generally. In the bacterial world the same is true for one photosynthesizing group – Cyanobacteria – but not for the others. Two of these 'others' are the green and purple sulphur bacteria. These groups are only distantly related to each other in the great bacterial branch of life, but they have a common feature that is reflected in their names – they make sulphur.

Recall that in land-plant photosynthesis water gets split into hydrogen and oxygen, and the oxygen is released into the atmosphere. The two groups of sulphur bacteria split hydrogen sulphide instead of water. The structure of the hydrogen sulphide molecule is $H_2S$, somewhat like water ($H_2O$). The splitting of $H_2S$ naturally yields not oxygen but sulphur, which is deposited inside or just outside the bacteria concerned. This type of photosynthesis is called anoxygenic, because no oxygen is produced.

The colours of the organisms belonging to the two groups of sulphur bacteria are determined by their photosynthetic pigments. Both groups contain a bacterial version of chlorophyll, which is simply called bacteriochlorophyll, though this is perhaps a good point at which to acknowledge that both this pigment and the chlorophyll found in plants are actually families of related molecules rather than just a single type. Green and purple sulphur bacteria have different bacteriochlorophylls; purple sulphur bacteria also have pigments belonging to the carotenoid group, which is why they end up looking purple.

The ecology of sulphur bacteria is interesting, and may provide a clue about how most bacteria (and archaea) survived in ancient environments on Earth, say three billion years ago, before the Great Oxygenation Event in which oxygen began to accumulate in the atmosphere. Present-day sulphur bacteria are typically found in environments that are oxygen-poor (hypoxic) or lacking in oxygen altogether (anoxic). These include the bottoms of lakes in which there is little mixing between the different layers of water, stagnant ponds, swamps, waterlogged soils, and sediments at the bottom of oceans.

Originally, all habitats on Earth were severely hypoxic or anoxic. In these ancient habitats (and in their more spatially restricted modern equivalents) the metabolism of organisms was anaerobic (not using oxygen), as opposed to aerobic (using oxygen). Although we're more familiar with the latter type of metabolism, as it's the norm in the animal and plant kingdoms, anaerobic metabolism is still common in microbes, both bacteria and archaea.

In the course of evolution, aerobic metabolism must have evolved from its anaerobic counterpart, since the use of oxygen prior to the Great Oxygenation Event would have been severely restricted. However, it's interesting that most of today's members of the group which powered that event – the Cyanobacteria – are aerobic. The resolution of this apparent paradox is probably that the original Cyanobacteria were anaerobic, while many of their descendants became aerobic, and some of them became able to switch between the two types of respiration. The details of these evolutionary transitions are not well established, though it seems likely that aerobic respiration has evolved independently several times in different lineages of Cyanobacteria.

Returning to the green sulphur bacteria, some of these do something quite remarkable. They live in the vicinity of deep-sea hydrothermal vents where there is no sunlight at all; yet they photosynthesize. How is this possible? It seems that they use what might be called geothermal radiation from vent fluid. Although this radiation is mostly infrared, it extends into the visible part of the spectrum. And it seems to be just sufficient in amount to sustain a low rate of photosynthesis. Another possible source of light for photosynthesis is bioluminescence. Again the amount is tiny, but there might be just enough of it, coming from various creatures of the vent community, to enable low-level photosynthesis. However, that said, the most important form of autotrophy in hydrothermal-vent communities is not photosynthesis but *chemo*synthesis.

## Chemicals Instead of Light

As we've seen, most photosynthesis involves the production of glucose from carbon dioxide and water using light energy. The carbon of carbon dioxide is said to be 'fixed' in the sense that it's transferred from inorganic to organic form. Energy is required to do this. But it doesn't have to be light energy; any energy that can be harnessed will do. And in hydrothermal-vent ecosystems, chemical energy is easier to come by than light. Since these communities are the basis for our discussion of chemosynthesis, let's have a general look at them before getting into the details. We'll start with the mid-Atlantic Ridge, a deep-sea location that we met in the previous chapter.

In fact, this is just one out of many ocean ridges. Others are the East Pacific Rise and the Southwest Indian Ridge. What they have in common is that they are all at creative plate boundaries – junctions at which the plates concerned are diverging and new crustal material is being formed by upwelling from the Earth's mantle. Moreover, they all join up, creating a worldwide 'web' of such boundaries/ridges. In total, this connected-up ridge is about 80,000 kilometres (50,000 miles) long, making it the Earth's longest mountain chain.

In cross-section, the topography of the boundary often takes the form of a split ridge with a deep rift in its centre. How pronounced this pattern is depends on the rate of upwelling and the speed of divergence: the slower the divergence the wider the rift. The hydrothermal vents are found in and around the rift and fall into two categories – black and white smokers. Black smokers are found in the more central, hotter regions (temperatures from 50 to 200 °C); these are flanked by white smokers in cooler regions (temperatures from 30 to 50 °C).

It's clear from the above that hydrothermal vents are not rare structures but oft-repeated ones spread along a very extensive system of ocean ridges. And everywhere we find vents, we find vent communities of organisms. We'll look at some animal members of these communities in the next section. Here, we'll focus on the microbial components without whose chemosynthesis there would be little in the way of organic compounds for the animals to eat.

The water that is spewed out of hydrothermal vents is high in nutrients. It has both oceanic and magmatic inputs, these being initially at very different temperatures. The oceanic input has been sucked down through fissures to either side of the vent and is then recirculated up through the vent itself. Before being sucked down, its temperature is just above zero degrees Celsius. The magmatic input has a temperature that can be anything up to about 500 degrees. When the mixture emerges at the vent and contacts ordinary seawater, some minerals precipitate out. These contribute to the building of the tall chimneys that often form at such vent sites. However, many minerals remain dissolved in the water, making it a rich broth that can be utilized by both bacteria and archaea.

Various chemicals dissolved in the water can be used as a source of energy. One of these is hydrogen sulphide gas, which we've already met (in the previous section) because it is split in anoxygenic photosynthesis. In chemosynthesis by archaea and bacteria, hydrogen sulphide is also used but in a different way – it's used as a source of energy to turn the carbon of $CO_2$ into organic carbon, a highly simplified version of the reaction being: $H_2S + CO_2 \rightarrow$ glucose + sulphur. Here, the arrow does not imply the use of light; rather, this is solely chemical energy, obtained from the oxidation of the hydrogen sulphide.

Although both archaea and bacteria can conduct both this and other forms of chemosynthesis, it's worth recalling that they are only very distantly related to each other. The domain Archaea is thought to be more closely related to our own domain, the Eukarya, than to the bacterial domain. And despite there being oxygen-generating photosynthesis in some bacteria (Cyanobacteria), it seems that there are *no* archaea that are capable of this form of photosynthesis. There are, however, some archaea that use light to pump protons (hydrogen ions) out of the cell, a process that ultimately drives the production of the energy-storing molecule ATP that is also produced by 'normal' photosynthesis.

## Energy in Ecosystems

Many animals have adapted to the extreme conditions that prevail in the vicinity of hydrothermal vents, including various groups of crustaceans, molluscs, and annelid worms. These are all ultimately dependent on the chemosynthesis carried out by bacteria and archaea. Some of the animals directly graze on the microbes, for example amphipod crustaceans that are related to the sand-hoppers we sometimes see jumping on beaches. Other animals, including their crustacean cousins the squat lobsters, eat the amphipods. If it were only these three components – chemosynthetic microbes, amphipods, and squat lobsters – that made up the ecosystem, we could refer to the energy transfer that occurs there as a food chain. Of course, neither vent systems nor any other ecosystems are as simple as this, and we are much better to think of an overall food *web* as representing the pattern of energy flow. But before making the transition from chain to web let's focus on the efficiency of energy flow in a chain.

The chain we started with – microbe–amphipod–lobster – is just one example of a chain involving an autotroph, a primary consumer, and a secondary consumer. Another simplified example is grass–impala–lion. In this case we would use 'herbivore' for the primary consumer and 'carnivore' (or predator) for the secondary consumer. An example of an avian food chain rather than a mammalian one is seed–greenfinch–sparrowhawk.

One thing that is reasonably consistent about food chains – regardless of their exact composition and whether they are marine, freshwater, or terrestrial – is that the efficiency of energy flow from one level to another rarely exceeds 10%. This limitation in the transfer of biological energy in food chains means that the number of links in any such chain is limited to about five. In the above examples there were just three. Lions have no predators, but small birds of prey like sparrowhawks are occasionally eaten by larger ones such as peregrine falcons, thus providing a fourth link. And a parasite of such a falcon could be seen as a fifth link. There are a few cases where six links can be found, but arguably none with seven, and certainly none with ten or more. Since this energy-efficiency limitation is independent of the groups of animals concerned, it is likely to apply also to food chains on exoplanets.

Now it's time to ditch the simplified concept of a chain and replace it with the more realistic notion of a web. None of the examples I've given of carnivores in the above account eat only the type of food I've mentioned. Lions eat a wide variety of prey, as do sparrowhawks and lobsters. And typically their many species of prey animals also consume many species of autotrophs. Indeed, many species of animals in ecosystems eat a mixture of autotrophs and heterotrophs. For example, blackbirds eat seeds, but they also eat worms. So the whole idea of a neat series of trophic levels, starting with autotrophs (level 1) and ending with whatever animal is at the top (say level 5) is flawed.

This conclusion, however, does not alter the fact that there is a general limit to energy transfer in ecosystems – now thought of as webs. At each transfer, much energy is lost as heat. Also, at most transfer points, material is unconsumed. The remains of pigeons left after sparrowhawk strikes in my garden can be considerable. And lions leave parts of impala carcasses uneaten, especially bones. Even when a prey organism is eaten whole, as in some cases by snakes, not all of the prey's body is digested. Some is egested as faecal material, which is then consumed by detritivores and decomposers

In a typical natural ecosystem, the number of species involved in a food web is huge. Rather than being just a few, it is typically a few hundred, and maybe in some cases – for example tropical rainforests – a few thousand. Some species have larger effects on the pattern of energy flow than others. The most important in this respect are called keystone species. One of the classic pieces of work leading to this concept was carried out by the American ecologist Robert Paine in his long-term studies of rocky intertidal food webs. He found that removing a starfish – the top predator in the system concerned – led to one of its prey species, mussels, outcompeting many other sessile animals growing on the same rocks; hence much more energy was channelled through the mussel population than when the starfish were present.

Further field experiments proved that major effects of removing certain species from a food web are common, though they are not always predictable in the form they take. And in general we would expect that not all species contribute equally to a food web's stability or energy flow. A typical food web has both major and minor players in the game of energy acquisition, and indeed there is usually a continuum from one to the other.

## Energy and Entropy

One notion that we didn't use, but perhaps could have, when defining life in Chapter 1, is that of entropy. This is a hard-to-define idea that belongs to physics (and more specifically to the field of thermodynamics), but it is basically the degree of disorder in a system. So, it can be thought of as the opposite of order. The more ordered a system is, the less entropy it has, and vice versa. According to the second law of thermodynamics, the degree of entropy (or disorder) of a system tends to increase over time. However, this law applies only to 'isolated systems'. These are systems in which energy and matter can flow within the system but not between it and other systems. Hence the universe is an isolated system (we think) but the biosphere is not.

It is because the biosphere is not an isolated system, and in particular because it is a system with continuous energy flow through it from an external source – the Sun – that it is characterized by living systems that exhibit many forms of increasing order (and hence decreasing entropy). These forms of ordering include evolution, embryonic development, and ecological succession. The last of these is the process by which a mature ecosystem becomes established in a place where it was initially absent, such as a newly formed volcanic island.

Succession from a starting point of bare rock typically begins with microbes and then some larger organisms such as algae and lichens. The latter are strange hybrid organisms that represent a form of symbiosis between kingdoms in that each lichen has a fungal and an algal component. Indeed, some lichens represent symbiosis between domains, because although in many lichens the photosynthetic partner of the fungus is a green alga (in other words a plant), in other lichens this partner is a species of Cyanobacteria (which used to be called blue-green algae). The existence of lichens further complicates the idea of food webs, because they are like the Venus flytrap with which this chapter started, in that an individual lichen 'plant' is both autotrophic and heterotrophic. And lichens are no minor component of the biosphere – there are more than 10,000 'species' of them. As an aside, it's worth noting that for practical purposes, lichenologists use the fungal component to assign a species name.

It is reasonable to expect that biospheres on other planets are characterized by energy-driven, entropy-decreasing processes that are similar to those observed on Earth – though of course this statement inevitably raises the question 'how similar is similar?' Perhaps other biospheres are also characterized by similar habitats, at least in terms of the broad habitat types, such as terrestrial and aquatic. We'll now look at Earthly habitats in more detail, and we'll try to keep alien biospheres at the back of our minds as we do so.

# 6   HABITATS AND LIFE

## Places to Live

We all know roughly what a habitat is – it's a place to live that has certain characteristics. It may be a patch of tropical forest, a pond, or the vicinity of a hydrothermal vent at the bottom of an ocean. And the general idea of 'habitat as home' seems almost certain to apply wherever we find life, albeit the kinds of habitats that exist on exoplanets are hard to predict. Life-forms need to live in some particular place, and it's reasonable to assume that any life-form will have some limitations to the conditions that it can withstand, and thus to the habitats in which it can survive.

However, while the notion of a habitat may be simple at one level, there are many subtleties attached to it at another. Some of these are associated with the sizes of organisms; some concern the distinction between habitats and microhabitats; some are to do with movement from one habitat to another during a single lifespan; and some are related to the issue of what organisms are doing in their habitat – it's easy for two types of creature to be found in the same place and yet have little or no ecological interaction. Let's now take a closer look at these subtleties.

We'll start by considering two very different kinds of animal that share a habitat: one species each of owls and tardigrades in a patch of deciduous woodland. We met tardigrades (or 'water bears') earlier, in connection with their ability to survive extreme conditions, including temperatures near absolute zero and the vacuum of space. Strangely, despite such amazing

capacity to withstand these conditions, many species of tardigrades live in very ordinary places. If we were to scrape the moss
off one of the trees in our patch of deciduous woodland and
examine its inhabitants with a microscope, we'd find many
tardigrades – typically 100+ in just a handful of moss.

Owls, as far as we know, are blissfully ignorant of these tiny
denizens of their woodland habitat. It's possible to get a good
idea of an owl's diet by breaking apart one of the pellets it egests
after eating, and identifying the bones that are found within it.
This reveals that for the most part the owl eats a mixture of
small birds and small mammals. But size is a relative thing.
These 'small' creatures are giants from the perspective of a
tardigrade that's less than a millimetre in length. The difference
is so great that not only do owls not interact with tardigrades,
but neither do their main prey organisms, for example mice. If
there's any interaction at all between the vertebrates and tardigrades of the wood, it's in a sense unintentional – perhaps a
small mammal might accidentally eat some tardigrades when
grazing on or near a patch of moss.

It's a good idea to make a digression at this point into a
different habitat – the open ocean, and in particular its upper
layers, which are rich in plankton. A major predator of plankton
is the shrimp-like crustacean called krill (this word is both
singular and plural, like 'sheep'). Krill are small animals – many
are about a centimetre in length, so perhaps 10 times as long as a
typical tardigrade. But some of their predators – for example
blue whales – are massive. So the rationale used above in terms
of size difference precluding ecological interaction is false – or at
least restricted to certain types of organism in certain types of
habitat.

The reason why no large predator has evolved to hoover up
terrestrial tardigrades in the way that blue whales have evolved
to hoover up marine krill is connected with the tardigrades'
microhabitat. If a predator tries to indiscriminately eat all organisms that live along a stretch of tree branch, it will obtain a diet
that consists mostly of moss, lichen, and bits of broken bark. The
tardigrades, though numerous, will be only a small proportion

of the total; also, rather different digestive processes are required for moss and small invertebrate food. In contrast, a whale can happily hoover up krill without getting a mouthful of relatively indigestible vegetable matter. It may well get a mouthful of plankton – the food of the krill – but this is not a problem. The smaller planktonic organisms will generally get washed out of the whale's mouth as it sieves the outgoing water through its baleen plates. Larger planktonic forms may be retained, such as the crustaceans called copepods, but these simply add a bit to the nutritional value of the catch.

Returning to our deciduous wood, the moss on the tree trunks can be said to be the microhabitat of certain species of tardigrade, while the habitat – or macrohabitat – of these creatures is the wood itself. But this distinction between habitat and microhabitat is harder to make for the owl, since it traverses much of the wood on its nightly flights. However, owl eggs and chicks most certainly do have a microhabitat – the nest. So, here we see a link between habitat and life stage. Let's now explore this further in the setting of a pond.

Dragonfly larvae are voracious predators in a typical pond. They attack and consume a wide variety of animal prey, including tadpoles. They may spend a considerable time in the pond – it depends on both species and latitude – in some cases about a year, in others two or more. But however long the time they spend there, this habitat only applies to part of the dragonfly's life cycle. Eventually, those larvae that survive long enough will emerge from the pond, climb onto a suitable leaf, and transform into an adult. This transformed creature has an entirely terrestrial existence – until one day it returns to the pond to lay eggs.

Significant changes in habitat during a lifespan are not always linked to 'complex' life cycles – those with larvae that are very different from the adults of the same animal. Migratory birds spend part of the year in one habitat, another part of the year in another – and this second habitat may be in a different continent than the first. For example, the barn swallow spends the summer in Europe, the winter in Africa. In this case it's not so much the *type* of habitat that differs with time – both European and African habitats of the swallow are open rather than densely forested areas – but rather the prevailing temperature. However, that said, the range of prey items available is very different in the two places. This is because most of the swallows' insect prey species are non-migratory. So a barn swallow in an African savannah in December has a very different diet to that same swallow when it reaches an area of European farmland the following April.

## Ways of Living

Discussion of diet takes us to another important distinction – that between habitat/microhabitat on the one hand and ecological niche on the other. Unfortunately, although 'niche' is a much-used term in ecology, it has been defined in many different ways. We'll look at two of these here to get an idea of the variation in usage.

The version of the niche that I like best is the one deriving from the work of the British ecologist Charles Elton. He pictured the niche of an animal as its role within its ecological community. He famously said, in his book *Animal Ecology*, 'When an ecologist says "there goes a badger", he should include in his thoughts some definite idea of the animal's place in the community to which it belongs, just as if he had said, "there goes the vicar".' In case you're not familiar with 'vicar', it's the Anglican Church version of 'priest'. Elton adds that an animal's niche is its relation to food and enemies, so its niche in this sense is its position in the food web rather than, for example, its physical position in the habitat. So the Eltonian niche, as it's sometimes called, encapsulates a way, rather than a place, of living. A badger's Eltonian niche might be described as 'large nocturnal generalist consumer, with a particular fondness for earthworms'.

Now contrast Elton's niche with that of the British-American ecologist George Evelyn Hutchinson. Elton's concept was informal, Hutchinson's formal. Here's what it is: an *n*-dimensional hypervolume bounded by the tolerance limits of the species concerned. But what does that mean, exactly? Let's see, and we can use $n = 2$ so that we can draw a graphical picture of a niche, with one of the dimensions being horizontal, the other vertical. Suppose that we're dealing with a seed-eating bird. But forget its food for the moment, and focus on its climatic tolerance. Imagine that one axis of the niche of this bird is temperature and another is relative humidity. On each of these axes (or dimensions) we can in theory pinpoint limits of the bird's tolerance. By joining up all four of these – maximum and minimum tolerable temperatures and humidities – we define a two-dimensional space.

There are many problems with this concept. For example, the temperature experienced by a bird living in a particular habitat will vary in many ways – daily, seasonally, and quasi-randomly as the weather fluctuates. The bird's tolerance to temperature might be partly related to its average and partly to its extremes. The same might be true of humidity. And if we move from two dimensions to many – for example we could add in dietary seed sizes as a third variable, and so on – the same problem of variation will apply to most if not all of the others. Not only that, but the variables may interact – for example maximum tolerable temperature may depend on humidity (Figure 6).

Since one or more measures of the diet can be axes in the Hutchinsonian niche, and since Elton's niche is largely diet-based, the two concepts overlap to a degree. Nevertheless, their differences outweigh their similarities. So a strategy that has been adopted by many ecologists is that in practical studies we should focus on something more restricted than an overall niche (in either sense), and refer simply to whatever variable we're focusing on, be it dietary, temperature-related, or whatever. That said, 'niche' is still useful in a general sense to refer to the idea of how rather than where a creature lives. Mode and place are related, but they're not the same; for example, you cannot eat canopy-level foliage if you're a ground-dwelling lizard.

It's hard to imagine an extraterrestrial biosphere in which the life-forms do not have some sort of habitats and niches – in other words, they must live somewhere on or near their planet's surface and they must make their living in some particular

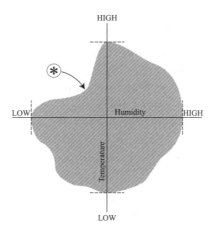

HIGH

LOW — Humidity — HIGH

Temperature

LOW

**Figure 6** The Hutchinsonian niche (shaded) of an unspecified bird species, shown in simplified form with just two dimensions: temperature and humidity. Maximum and minimum tolerances for each variable are shown as dashed lines. The two variables interact: for example, note the asterisked concavity at the top left, where we see that this species has a particular difficulty with conditions that are both hot and dry. It is easy to picture a third dimension, for example the range of seed sizes the bird can eat, projecting into and out of the page. Other dimensions can also be added, but this is harder to picture, as it involves a shift from ordinary (three-dimensional) space to hyperspace.

way. So it seems likely that these two concepts will transcend their Earthly origins. This is worth remembering, because in ecology we use familiar animal names like bird and badger, thus making the whole endeavour seem familiar and local, whereas in fact habitats and niches may be applicable across the Milky Way and beyond, wherever we find inhabited planets.

We'll probe further into this possibility in the final section of this chapter – though such probing will necessarily be of a rather general nature. However, for the Earth we can go into as much detail about such matters as we like. Detail is a funny thing – too much of it is boring, but too little of it can lead to glib overgeneralizations. So, in the next three sections we'll look at three very different habitats on our own planet, and will focus on particular features – and creatures – of each; and we'll try to steer a 'middle course' in terms of the amount of detail.

## A Patch of Ground

In the previous chapter, we discussed some simplified food chains, such as grass–impala–lion and seed–finch–hawk. However, such examples, though common in the ecological literature, are highly biased for the following reason. Most of the energy of plant material in a land-based ecosystem never enters the herbivore trophic level. Rather, it goes straight to detritivores. Most leaves of a maple tree do not get eaten by herbivorous insects – instead they fall from the tree at the end of the summer season and begin to decay. Parts of them will probably be consumed by earthworms. From there on, what happens to the energy is largely a subterranean affair. Out of sight, out of mind, as they say – hence our disproportionate focus on the bits of the food web that we can see. So let's set the record straight by considering the subterranean part of the functioning of a land-based ecosystem. Hence the name of this section: we're going to pick a patch of ground, and then go down rather than up.

Let's start in the same patch of deciduous woodland in which we saw owls eating small mammals and other prey at the start of this chapter. It's both a theoretical patch of woodland representing all such habitat and a particular patch of woodland in County Antrim that I used to visit with my brother when we were young.

Going down from the plants of the forest floor, there's a series of layers – they're all ill-defined, but they're layers nevertheless. Under the ground flora there's a layer of leaf litter. This is composed of the dead leaves of the previous autumn. This layer fluctuates in thickness both in time and space. It's thickest where leaves collect in hollows, and in the month of November. But it's rarely absent from the wood. The leaves of trees decay rather slowly, especially when dry, and many of the old leaves are still there on the ground when the new ones first appear as buds at the start of the next spring. Millipedes and centipedes are characteristic animals of the leaf litter; thus they share a microhabitat. However, it's worth noting that their niches are very different: millipedes are generally detritivorous, while centipedes are universally carnivorous.

Below the leaf litter is a layer where dead plant and animal matter is more thoroughly decayed – the humus. Further down, the humus then intergrades into the topsoil – a layer of soil that's rich in organic matter. The topsoil is an ecosystem in its own right. It has a high biodiversity of both animals and microbes. Although there are no strictly subterranean plants, there are major plant components of the topsoil in the form of roots. Indeed, roots themselves provide a sort of microhabitat within the topsoil layer. They're usually associated with fungi in a system called the rhizosphere. This use of the term 'sphere' is an even greater taking-of-liberties than in the case of the biosphere. As we saw in Chapter 4, the biosphere is approximately a spherical shell. But a rhizosphere is not even that – it's an irregular shape whose precise details depend on the plants whose roots form its basis in any one particular place.

Earthworms are a key component of the topsoil layer. They are major players in the acquisition of organic matter from the leaf litter above – they drag fragments of dead leaves into their burrows – and this is essential for the powering of the soil system. The importance of earthworms did not escape the attention of that great naturalist Charles Darwin, who published a book about them in 1881.

Earthworms are eaten by various predatory animals. Perhaps best known for having this ecological niche is the mole, whose underground galleries are patchily distributed, as we can see from the positions of their surface openings. In fact, they're commoner in open pasture habitats than in forests – and moles are entirely absent from Ireland, so there are no molehills near that patch of County Antrim woodland that my brother and I visited those many years ago.

But earthworms are not attacked only by moles. As we've already seen, badgers have a penchant for them. And they have invertebrate predators too. There's a little-known but ecologically important group of subterranean centipedes called the Geophilo-morpha. I've added hyphens to its name to break it up and to reveal more clearly its meaning – 'ground-loving forms'. There are many species within this group. Those of larger body size include worms in their range of prey items. Indeed, there are reports of some species of these centipedes 'pack-feeding' on larger earthworms. The centipedes use their venom claws to inject a mixture of toxic chemicals into their prey; then those same claws are used to rip holes in the worm's integument to facilitate entry of the centipede into its new-found larder.

Of course, the soil and the wood above it are not isolated from each other. Not only does energy go downward in the form of falling leaves, it also goes upward, for example when a blackbird wrenches an earthworm up out of the topsoil.

The next layer down is the subsoil. If you've ever dug a large hole to make a pond, you'll have seen the junction between these two layers very clearly. The dark brown of the topsoil gives way to the orangey-brown colour of a typical subsoil at a depth that's variable but often only a few spadefuls down from the surface. This lower layer is home to a less biodiverse ecological community than the topsoil above it, but there are still plenty of organisms per cubic spade – many of them bacteria. Below the subsoil is the bedrock, which can be thought of as the top of the Earth's crust. Usually, these two layers intergrade through a zone of broken-up bedrock interspersed with soil and grit. Again, there are many microbes here, and even a few small animals, but not many. The nature of the ecosystem this far down is very different from its topsoil counterpart.

## In the Ocean Depths

Let's now revisit another old haunt – the hydrothermal vents of a deep-ocean ridge such as that of the mid-Atlantic. When we first encountered these in Chapter 5, we focused on their source of biological energy – chemosynthesis – and the microbes that are responsible for it, briefly mentioning a few types of animal that utilize this energy. Now let's look more broadly at the vicinity of these vents as a habitat for many weird and wonderful animals.

Most of these animals belong to just three of the more than thirty phyla of the animal kingdom – arthropods, molluscs, and annelids (segmented worms). Vent arthropods are all crust-aceans of one sort or another, while most vent molluscs are snails, bivalves, or cephalopods, and vent annelids mostly belong to the group called beard worms. All of these are extremophiles that have to endure conditions that would be alien to the animals of the surface waters (e.g. plankton), which would die in a very short time if transported to a vent. Both temperature and pressure are extreme near hydrothermal vents, though in a way this combination is helpful. At the surface of the Earth, where the pressure is described as being 'one atmosphere', water is in gaseous form if the temperature is above 100 degrees Cel-sius – as in the case of a boiling kettle. But if the pressure is ten times higher, water remains in liquid form until the tempera-ture is well above 110 degrees.

At hydrothermal vents, the temperature can be more than 300 degrees Celsius and the pressure more than 200 atmospheres. At the most extreme combinations of these vari-ables encountered around vents, water is in the form of a 'super-critical fluid' – a strange phase that has some of the properties of a liquid and some of those of a gas. But most vents seem to be characterized by water in its liquid form. The situation is com-plicated by the fact that the salinity level of the water affects its phase transitions, and salinity can be quite variable in the areas around vents.

Vent animals can be sessile – attached to the vent chimneys or to other substrata nearby – or mobile. In the latter case, they can either walk on the surface of the substratum or swim above it. Beard worms – there are about a hundred known species of them – are all attached. They have no gut. But they have an organ called a trophosome, which is home to millions of chemo-synthetic bacteria; this is how they obtain food. The largest beard worms – giant tube worms of the genus *Riftia* – are longer than a human arm.

Vent molluscs may be attached (bivalves) or mobile (snails, cephalopods); some of the snails are slow-moving limpet-like forms. Vent crustaceans are likewise a mixture, with sessile barnacles and mobile shrimps. Since many groups of invertebrates are represented in vent communities, it's clear that evolution from 'normal' to extremophile forms has occurred independently in each of these. For example, vent shrimps are more closely related to their rock-pool cousins than they are to their vent-snail cohabitants. In other words, we are convinced that the first ever shrimps and snails on our planet did not originate near vents and then give rise to descendants that colonized rock-pools. However, note that the opposite direction of evolutionary change is also possible in other taxa, such as bacteria or archaea. Some scientists argue for the origin of life having taken place near vents, others for the origin of life occurring in the more benign surface waters of shallow seas at the edges of continents.

The nature of a vent food web varies with location, as does that of a forest. But here's a skeletal version that is broadly applicable to this kind of ecosystem. Various bacteria and archaea are the primary producers of the system, fixing carbon by chemosynthesis, using hydrogen sulphide or other inorganic molecules as their starting point. Some of these microbes occur as mats that are consumed by snails and other grazers, including shrimps and other small crustaceans. Others live in the tissues of beard worms, and indeed in the tissues of other vent animals too. Predatory forms that eat the grazers include crabs, elongate fish belonging to a family called cusk-eels, and at least one species of small extremophile octopus that eats vent crustaceans.

Regardless of the details of the food web in the vicinity of any particular black or white smoker, the existence of these deep-sea communities is testament to life's ability to thrive under conditions that would at first sight seem to preclude living organisms altogether. It is this fact that makes the subterranean oceans of some moons of Jupiter and Saturn such attractive targets for astrobiological missions of the future. However, that said, life originating in such places is not the same as life adapting to extreme conditions on Earth from a starting point of less weird life on our planet's surface. Neither Europa (moon of Jupiter) nor Enceladus (moon of Saturn) has a surface on which life as we know it would be possible; so there is no cosy environment on these rocky bodies that could serve as a source from which to evolve extremophile animals capable of surviving their ocean depths. On these moons the surface is an even more extreme environment, from the perspective of life, than the cryptic oceans that lurk beneath.

Although we're pretty sure that each group of vent animals on Earth evolved from ancestors dwelling near the sea surface, exactly how that evolutionary process occurred is far from clear. Living at a pressure of over 200 atmospheres is paradoxically not a problem, because the pressures within and outside an animal's body are the same. But how does evolution get from an ancestor accustomed to one atmosphere to a descendant accustomed to 100 times that pressure? The same goes for temperature, though perhaps the situation is facilitated here by considerable variation in temperature around the vents. The outflowing ultra-hot water meets seawater that is only a couple of degrees above zero Celsius. As the water mixes, a range of temperatures is produced, so mobile animals can seek out areas where the temperature is tolerable. But sessile animals cannot.

Studies on the Pompeii worm – another annelid like the giant tube worm but much smaller (about 10 cm in length) – have revealed a temperature gradient in the tube in which a worm lives from about 80 degrees Celsius at one end to about 20 at the other. This means that there is a temperature gradient of about 60 degrees Celsius along the worm's short tube. The orientation of the animal is head-out. In other words, its tail is in the part of the tube that's attached to the wall of the smoker, its head at the other end. This is perhaps more comfortable than the reverse arrangement, but still an incredible feat of survival. Enzymes have to work in all parts of the body of any animal – not just in the head – and they, like all proteins, are denatured by heat. Somehow, evolution has modified the proteins of the vent animals to enhance their thermostability way beyond the norm.

## Twixt Land and Sea

Of the two examples of habitats that we've looked at so far, the benign one – the woodland – is familiar to us from personal experience, while the extreme one – the hydrothermal vent – is not. However, this doesn't mean that all extreme environments are unfamiliar – far from it. Many of us spend holidays near the sea, and we tend to think of the coast as a place conducive to relaxation. But for the animals that live there it's not relaxing at all. Let's examine the most extreme zone of the coastal environment – the intertidal. We'll focus in particular on a rocky intertidal zone. Our conclusions will apply in a general way to sandy shores too, but the details will differ because the animals themselves are different.

A typical rocky shore can be divided into zones that differ in the relative amounts of time that each spends immersed (at the high end of the tidal cycle) and emersed (at the low end). The exact pattern of zonation of the shore differs from place to place, in response to various factors, including different tidal patterns and different amounts of spray. But the general trend from high to low shore is the same, with animals that are most resistant to desiccation, and/or least tolerant to being submerged, being found at higher levels.

Let's compare three types of animal found in rocky-shore systems: sea slaters, barnacles, and sea urchins. If we were to follow an outgoing tide down the shore, we'd find sea slaters at the top, barnacles in the middle, and sea urchins at the very bottom. However, as with all ecological generalizations, there are various complications, as we'll now see.

Although a sea slater and a barnacle are both crustaceans, the former breathes air, the latter breathes water (with a complication discussed below). The evolutionary history of a sea slater, in terms of habitat, goes as follows: marine → terrestrial → intertidal. In contrast, the evolutionary history of a barnacle goes: marine → intertidal. Sea slaters don't survive long when immersed. Equally, barnacles can't survive long periods of emersion (though surprisingly some species have been shown to be able to survive for days rather than just the few hours that are normal in the tidal cycle). However, one complication with this comparison is that while the common name 'sea slater' refers to just a few species (genus *Ligia*), the name 'barnacle' refers to many species within a large group. This is a group about which Charles Darwin wrote several monographs, and indeed his interest in barnacles is itself the subject of a 2004 popular science book by Rebecca Stott – *Darwin and the Barnacle*. The barnacles I had in mind when writing the above comparison are intertidal ones belonging to the family Balanidae.

A similar complication applies with sea urchins. The many species of urchin span a vast range of depths. Many are not intertidal at all, but rather subtidal. However, a few species may be found at the bottom of the intertidal zone. And at least one species – the shingle urchin (*Colobocentrotus atratus*) that's found in Hawaiian waters – has evolved to be able to withstand spending much of its time in the intertidal. An adaptation that helps it in this respect is its shape – it is much squatter than most urchins, which means that it is more resistant to being dislodged from the substratum by waves.

Another complication to the above comparisons is movement. It's only possible to speak of zonation if the species concerned are sessile or at least of restricted mobility. And the degree to which these kinds of animal are mobile varies considerably. Sea slaters are highly mobile, like their inland woodlouse cousins, so their restriction to the top part of the shore is due to an active choice, on their part, to avoid the risks of going too far down the shore to obtain food. If we could illustrate their zonation with a map of their abundance at any moment in time, we'd find a high density of sea slaters at the top of the shore, a lower density halfway down, and a very low density (at low tide only) near the bottom.

In contrast, intertidal barnacles are effectively glued to the rock, so once attached they can't shift their positions at all. About a decade ago, the mechanism of their glue was revealed – interestingly, it has some properties in common with human blood clotting. But what lifestyle does a barnacle have prior to becoming affixed to one spot? In fact, barnacle larvae are highly mobile. They are moved passively by water currents and tides, and they also have appendages that can be used to actively swim. It is the final larval stage that decides where to stop and turn itself into a sessile adult.

Sea urchins are different again in terms of movement. Like barnacles, they have mobile larval stages. But unlike barnacles, the adults can also move, albeit rather slowly. They move on their many tube-feet, which are similar to those of their fellow echinoderms, the sea stars (or starfish).

Given these various patterns of movement, immobile adult barnacles best illustrate the problems of the intertidal zone, which is an extreme environment in a different way than is the vent system that we looked at earlier. The areas around vents have some characteristics that don't vary much in space or time – notably their high pressures. The intertidal is extreme not so much in the actual values of the variables found there, but rather in the pattern of variation itself. The problem for an intertidal animal – especially an immobile one – is that its habitat alternates between aquatic and terrestrial on a very short timescale.

An adult barnacle is adapted to dealing with desiccation during its habitat's dry phase by having hard plates that it can close, thus almost sealing off the humid environment inside its shell from the more hostile, drier environment outside. However, the 'almost' is important because complete protection from desiccation would mean a complete block to incoming oxygen. So leaving a small opening is a compromise between these two conflicting demands. Regarding the acquisition of oxygen, barnacles are unusual as marine crustaceans in that they lack gills. They absorb oxygen from the water by diffusion. It has been calculated that the amount of water left within a barnacle's shell after the tide has gone out is only sufficient to allow it to survive for a few minutes. Thereafter, and until the tide comes in again, it has to cope either by absorbing oxygen from the air or by metabolizing anaerobically. Various studies suggest that barnacles primarily do the former.

In this chapter, we've seen certain advantages of animal movement, such as the ability of swallows to change continents with the seasons and the ability of sea slaters to roam the intertidal zone at low tide. Movement is also thought to be conducive to the evolution of intelligence – a subject that will be our focus in Chapter 8. It's also linked to skeletons, which is where we're going in Chapter 7. Just one final point before leaving the intertidal: the main cause of tides on Earth is the Moon. If we had no Moon, the habitat where the land meets the sea would be a very different place. And more generally, the presence or absence of moons has been hypothesized to be one factor influencing a planet's habitability, as will be discussed in Chapter 11.

## Parallel Places

Our final task in the current chapter is to think about the possibility of habitats existing on exoplanets that are broadly parallel to those here on Earth. This is an important issue because habitats have a major effect on evolution – and are in turn influenced by, and even created by, the evolutionary process. For example, forest habitats arose from the evolution of trees, and they have had a major influence on the evolution of the animals that inhabit them.

Planets in the habitable zone, especially those that are actually inhabited, are likely to have liquid water on their surfaces. Thus there are aquatic habitats and – except in the case of 'waterworlds' – land-based habitats also. Subdivisions of these two main habitat types are almost certain to exist, wherever they are found. Although the extent to which they parallel subdivisions here on Earth is hard to envisage, some broad similarities are to be expected. All rocky planets must have some sort of topography – the spherical equivalent of a bowling green is not likely to exist in nature at a planetary scale. Given that there will usually be hills, mountains, plains, and valleys, as we see on Mars, water will run downhill under the ubiquitous influence of gravity. It will tend to pool in depressions, whatever their size. Thus the equivalents of ponds, lakes, and oceans are likely rather than just possible. Indeed, we see such features on Titan (moon of Saturn), though in that case the liquid that pools is not water. I call the idea that habitable exoplanets typically have broad habitat types similar to those of the Earth the Parallel Places hypothesis.

On land, there will be equivalents of warmer, low-altitude habitats and cooler, high-altitude ones, again due to the omnipresence of topography. And there will be habitats at the edge of the land with special features, though whether these include significant tides depends, as we saw earlier, on the presence or absence of moons. The exact distribution of areas of water and emergent land will influence the extent to which latitudinal variation will produce warm and cold seas and landmasses, but *any* distribution will have some effect of this kind. And indeed, as we saw in Chapter 4, the distribution of oceans and continents has changed markedly over geological time on the Earth.

The most general types of habitat, such as those of different altitudes, originate through physical processes such as mountain-building. More specific habitat types are to a large extent created by evolution itself. For example, arboreal habitats for animals such as primates are only possible where there are trees. Are trees a usual outcome of the evolutionary process on an Earth-like planet, or is their existence vanishingly improbable? Given that a tree-type growth-form has been hit upon independently by different lineages on Earth (tree-ferns, conifers, many families of flowering trees), it may be a commonplace feature of biospheres elsewhere. If it is, then we should expect arboreal animals to be common too. This is of particular interest from the perspective of evolving creatures resembling apes – but only if there are equivalents of vertebrates elsewhere to provide the skeleton that can be adapted to climb trees.

# 7 SKELETONS AND LIFE

## The Need for Support

Many environmental factors have complex patterns of variation across the biomes of the world, but one of them does not: gravity. All organisms need some means of dealing with the gravitational pull that tends to draw everything down to Earth. That means they need some sort of skeleton. Some skeletons are obvious, like our own, while others, such as those of worms and sponges, are not. All are essential for survival and, other things being equal, the bigger the organism the more important is its skeleton.

But other things are not always equal. The most important inequality in this respect is that between land and water. Note that the starting point above was 'biomes', and recall from Chapter 4 that these categories – including rainforests, savannah, and tundra – are used only for the terrestrial part of the biosphere. In the World Ocean – the phrase we sometimes use to refer to the system of all oceans that dominates our planet's surface – there are no direct equivalents of biomes. Rather, the emphasis is on interconnectedness and on the fact that some conditions do not vary as much as they do on land. The most obvious example of such a condition is the availability of water. It's not a problem for organisms living in a rainforest to acquire water, but it certainly is a problem in a desert. In contrast, one ocean is much the same as another in this respect.

The important thing for us right now, though, is not the variation or lack of it between one land-based system and another, or between one marine system and another, but rather

the difference between land and sea. And although this difference has a big effect on the need for support, it is not a difference
in the force of gravity, but rather in the force that counteracts it,
namely buoyancy. This upward pressure is much greater in
water than in air. And it's greater in the sea than in a lake. It's
particularly great in the Dead Sea, whose salinity is about ten
times that of an average ocean. In general, the higher the salt
content of a body of water, the greater is its buoyancy.

The difference in buoyancy between air and water explains
why it is possible for an animal to be very large without a
significant skeleton in the sea but not on land. The giant squid
can grow to a length of more than 10 metres. And there's
another species – from a different family – called the colossal
squid, which is similar in length but somewhat heavier. Squid
are molluscs, but they have largely lost their shell. They have
only a remnant of the original mollusc shell – it is a small rod-
shaped internal structure called a pen.

Now consider the possibility of a land animal of a metre or
more in height without hard parts. There aren't any. The reason
for this is that the downward pull of gravity is too great, when
unopposed by the buoyancy of surrounding water, to allow such
an animal to retain its shape. What gravity is 'trying' to do can
be seen using a liquid, for example wine. If you keep your glass
upright, the depth of wine can be considerable; this is particularly clear in those (unwise) glasses designed to hold an entire
bottle's worth. But if you accidentally tip your glass over (i.e.
remove the 'skeleton'), its height becomes only a few millimetres at most, gradually tending towards zero.

Inside the cells of all organisms on Earth is the fluid we call
cytoplasm. As with wine, gravity tends to pull this to the ground
if it's unsupported. But it is supported – by cell membranes and
other microstructures that we'll look at in the next section. And
in all large land animals the micro-level support is supplemented
by the macro-level support of bones or other hard structures.
The same applies in plants. Tall land plants – trees – have
'skeletons' made of wood, regardless of whether they are broadleaves, conifers, or tree-ferns. In contrast, the brown algal forms

that make up the kelp forests of some shallow seas can reach heights of many metres regardless of their lack of a hard skeleton. Admittedly they're not true plants, but in this case the taxonomy is unimportant. Gravity treats all organisms as equals.

The tallest trees – redwoods – can reach heights of more than 100 metres. The tallest land animals today are giraffes, some of which are over 6 metres. The tallest animals of the past are dinosaurs of the group Sauropoda, some of which were twice the maximum height of a giraffe. These great heights attest to the power of hard structures such as wood and bone to enable organisms to counteract the downward pull of gravity.

## Cellular Micro-Skeletons

We've already noted the importance of the cell membrane in keeping the cytoplasm contained. Now we need to venture inside the cell to look at a network of long thin structures that are collectively called the cytoskeleton. What follows is based on the eukaryotic cell that is characteristic of animals, plants, and fungi. Prokaryotic cells (as in bacteria) also have a cytoskeleton; it is similar to the eukaryotic one in general terms but differs somewhat in detail. Of course, there are also differences in detail between different eukaryote groups, and between different cell types within any one group; but we will ignore these here, as our focus is on the general principles.

There are three main components of the cytoskeleton: microtubules, microfilaments, and intermediate filaments. The differences between them include both their diameters and the proteins of which they are made. In terms of diameter, microtubules are thickest, microfilaments thinnest, and intermediate filaments in between, as their name suggests. At this level of microstructure, diameters (and other lengths) are measured in units of nanometres (nm), with one of these being equal to a billionth of a metre. The approximate diameters are 25, 7, and 10 nanometres.

Together, these three components of the cytoskeleton help the cell to retain its shape – at least over those periods of time when shape stability is called for. At other times, when shape change or even cell movement is called for, the cytoskeleton also helps. In both these respects this mini-skeleton is doing a parallel job to its macro-skeletal counterpart. For example, our bony human skeleton can help the body to remain upright for long periods, but it also plays an essential role in the everyday processes of sitting down or lying. So the skeleton of the cell and that of the body have parallel functions.

However, the cytoskeleton is better at multitasking than its macro counterpart. It has other functions than helping to maintain – or change – cell shape. For example, much within-cell transport of various materials uses the cytoskeleton as a system of pathways along which molecules and larger things such as organelles can be relocated from one position in a cell to another. And when a cell divides and one set of chromosomes is dragged into one daughter cell and the complementary set into the other, it is the microtubules that achieve this dragging. Naturally, during this process some of the microtubules have one end attached to a chromosome.

Although the cytoskeleton is better than its large-scale equivalent at multitasking, the macro-skeleton is not solely used for movement and stasis. Think, for example, of the bone marrow, where, deep inside a particular component of the human skeleton – for example the tibia of the lower leg – blood cells begin their lives.

Another way in which the cell's and the body's skeletons differ is in their constancy over time. Nothing in a living organism is entirely constant over time, but bones are much more constant than, for example, microtubules. My tibia hasn't changed much over the last decade, it's just a little bit 'the worse for wear'. In contrast, my cells' microtubules are being constructed and destructed all the time. So the cytoskeleton is in a state of continual flux. However, this flux does not detract from the skeletal function of the material concerned.

In many groups of organisms, as well as there being a mini-skeleton within the cell, there is another mini-skeleton outside it. What I'm referring to here is the cell wall. Although we animals generally don't have cell walls (we just have cell membranes), we're unusual in this respect. Other eukaryote groups – plants and fungi – have cell walls. So too do prokaryotes – both bacteria and archaea. Both the overall structure of the cell wall and the molecules it is made of vary between taxonomic groups. Nevertheless, in general it is much thicker than the cell membrane that lies immediately inside it. A typical cell wall has considerable strength, and yet is more flexible than its name suggests – it is not like a human-built wall in this respect.

## Armoured Exteriors

A little more than 520 million years ago, a type of animal appeared in the biosphere that would survive for more than half the time between then and now. This amazing survival machine was the trilobite – a marine arthropod whose armoured body is laterally divided into three lobes – hence its name. Of course, 'the' trilobite is shorthand for a group of thousands of species. These were highly successful animals in terms of diversification as well as overall timespan. If you want to learn more about them, there's an excellent book called *Trilobite!* written by British palaeontologist Richard Fortey, and published in the year 2000.

The closest living relatives of trilobites among today's fauna are the crustaceans, the insects, the arachnids, and the myriapods – in other words the four big groups of arthropods that have not yet gone extinct. Arthropods, both extant and extinct, are characterized by exoskeletons. However, these are very variable in their degree of rigidity. Some crustaceans such as crabs and lobsters have rather rigid exoskeletal plates. In other crustaceans, such as the terrestrial woodlice and the marine krill that we encountered in Chapter 6 as the staple food of some whales, have more pliable armour.

Within the other main groups of arthropods, there is also much variation. In the myriapods, some of the sausage-shaped millipedes that belong to the family Julidae have an exoskeleton made of rigid rings; in contrast, the subterranean centipedes we met when dealing with soil ecosystems have much more flexible armour. Likewise, among insects, beetles are more heavily armoured than flies. And in the arachnids, a scorpion typically has more rigid exoskeletal plates than a spider – regardless of whether the latter is a house spider or a tarantula.

In all the major groups of arthropods, the exoskeleton contains two main types of macromolecule: proteins (for example of the collagen group) and a nitrogen-containing carbohydrate called chitin. These can be thought of as forming a kind of semi-pliable meshwork. In those arthropods where the exoskeleton has become rigid (like lobsters), the hardening is achieved by a process called biomineralization. The main mineral that's incorporated into exoskeletons is calcium carbonate – which is abundant in seawater, and also in chalk and limestone. In fact, most limestone arises from the compaction of the biomineralized exoskeletons of arthropods, the shells of molluscs, and the hard parts of a variety of other organisms.

Arthropods in general, and the arthropod exoskeleton in particular, arose either in the early Cambrian period between about 540 and 520 million years ago, or earlier in the late pre-Cambrian. There is still much debate over this issue. And indeed the original arthropod may not have had an exoskeleton – that may have come along later. In general, the sudden appearance of animals in the fossil record that we call the Cambrian explosion is still in many ways an enigma.

Regardless of its origin, the exoskeleton's retention in so many groups of arthropods is testament to its utility. It's not without its problems, especially in terms of growth. Arthropods have to moult to grow, and in general these animals are more vulnerable during and immediately after moulting than they are at other times. But clearly the benefits outweigh the problems. For most arthropods, the exoskeleton is essential for the maintenance of body-form and – by providing multiple sites for muscle attachment – for movement. Think of the rapid movements of a praying mantis or of a mantis shrimp: these would be impossible without the exoskeleton. The same goes for the slower movements of a walking millipede or woodlouse.

But why have I referred to 'most' rather than 'all' arthropods in the above paragraph? Are there any arthropods in which the exoskeleton has become redundant for form and movement? Inasmuch as it's possible to generalize about a group that contains more than a million species, the answer seems to be 'no' with regard to *complete* redundancy, but 'yes' to partial.

There's a group of barnacles whose members are most unlike the 'normal' rock-encrusting ones. These parasitic forms – they parasitize crabs – have a bizarre body structure. The adult female consists of two parts, called an interna and an externa. The former is like a ramification of fungal hyphae that extends throughout the crab's body, while the latter is a bulbous sac-like structure that projects out of the crab's underside. There are no hard plates of exoskeleton in this animal, and it cannot move, unless we count the growth of the 'hyphae' as movement; though that said, the larvae are mobile, and closely resemble those of other barnacles.

Turning back to more normal arthropods with more normal exoskeletons, how large has evolution been able to make these creatures? We know of extinct dragonflies with wingspans of more than half a metre, and extinct millipedes that were more than 2 metres long. Extant Japanese spider crabs (crustaceans, not arachnids) can have leg-spans of more than 3 metres. With regard to height, some stick insects when standing upright are over half a metre tall. Although such sizes are dwarfed by vertebrates, they are still impressive – and they would not be possible without a skeleton made of hard parts. Invertebrates with other sorts of skeletons (which will be dealt with in the final section of this chapter) cannot achieve such leg-spans or heights – though some of them do achieve great lengths.

In terms of body size, vertebrates rule the animal kingdom; but in terms of species diversity, arthropods put us to shame. In contrast to the 50,000 or so extant vertebrate species, there are more than a million species of arthropods. And those are just the ones we know about and have named. So the evolution of an exoskeleton was one key to success in terms of animal body plans, while the evolution of an endoskeleton in the vertebrates was another. Apart from the arthropods, there's only a single animal phylum that has been more successful than the vertebrates in terms of known present-day species numbers. This is the phylum Mollusca.

## The Disposable Shell

Most molluscs have a hard mineralized structure (the shell) and it's usually – but not always – on the outside. However, 'most' is just that – the majority. Not all molluscs have shells; some have lost their shells as they evolved.

In the long course of animal evolutionary history, there is much opportunity to lose structures that have been gained if, in a particular ecological situation, they become disadvantageous, or if – and this is another version of the same thing – the benefits of having them become outweighed by the benefits of losing them. However, in some cases a structure becomes so deeply embedded in the overall design of a particular group of animals that it is almost never lost. We already saw that the arthropod exoskeleton has been downplayed in some cases (parasitic barnacles), but these constitute a tiny minority of arthropod lineages. And in vertebrates the endoskeleton has never been downplayed much, though of course when limbs are lost (as in snakes) the appendicular skeleton is lost too.

In this respect, a shell is a very different type of structure to suits of armour or bony interiors. Having been gained in the molluscan stem lineage, shells have been reduced or lost many times subsequently. We'll take a quick tour of the three main groups of extant molluscs – gastropods, bivalves, and cephalopods. But first let's look at the nature of the shell itself.

Like a lobster's exoskeleton, a mollusc's shell is built of two components, an organic one (made largely of proteins and carbohydrates) and an inorganic one. The latter, resulting again from biomineralization, is usually calcium carbonate, the same mineral used by the lobster. But there are some twists to this tale, and a particularly interesting one is found in a snail from the deep-sea vent communities that we looked at in the previous chapter.

This species is called the scaly-foot snail, and its shell is unusual in terms of its mineralization. Not only does it have the usual calcium carbonate, but it also has iron sulphide. This is linked to the unusual habitat: vent waters are rich in sulphides, whereas the habitats of the vast majority of the world's snails are not.

Let's now broaden out from one gastropod to all of them. The gastropods are by far the most successful of all molluscan groups in today's biosphere. There are about 80,000 species of them, compared with between 10,000 and 20,000 of the next most common group – the bivalves – and a mere 1000 or so of the 'brainy' group, the cephalopods.

It's a fairly safe bet that the first ever gastropod had a shell. It probably lived in the Cambrian period, about 520 million years ago, or possibly earlier, and it was a marine animal. The land wasn't colonized by any animals – molluscs or others – until much later, as we've seen. Of the thousands of lineages that this stem-gastropod gave rise to, most have retained their shells, but a significant number have not. Shell-less gastropods are generally referred to as slugs. However, since this term, when used in an unadorned way, most readily conjures up a mental image of a land-based creature, we use the term sea slug to refer to its marine equivalent.

The most numerous sea slugs in terms of species numbers are the nudibranchs (= 'naked gills'). There are about 3000 species of these. They typically retain a shell in their larval period and then lose it as they become adults. Many species of sea slugs are characterized by very beautiful adult forms – multicoloured, with laterally extended 'wings'. Their beauty is in marked contrast to the slimy ugliness that for most people is associated with land slugs.

The terms sea slug and nudibranch are sometimes used interchangeably, but this is incorrect, because some groups of marine gastropods that are not closely related to nudibranchs have independently evolved shell-less forms. And the same is true on land: different groups of land snails have given rise to land-slug descendants. So a particular family of slugs is often more closely related to a snail family than to another slug one.

The other two main groups of molluscs – the bivalves and the cephalopods – are in a sense opposites to each other in terms of shells. Bivalves – such as clams, oysters, and mussels – typically retain both their two shells (or valves), whereas cephalopods – including squid and octopus – generally have reduced or lost their (single) shell. However, at an earlier stage in their evolution, most cephalopods had rather impressive chambered shells, and at first glance look a bit like gastropods.

In fact, these extinct molluscs – grouped under the name ammonites (fossil-shop term) or ammonoids (biological term) – were dominant members of the marine fauna for more than 200 million years, before their demise in the same mass extinction event that killed off the dinosaurs about 65 mya. They ranged in size from a few millimetres across to more than 2 metres. The reduction or loss of shells in their squid and octopus cousins has been associated with an increase in mobility and speed, which we assume must have more than compensated for the loss of the protection that the shell afforded.

Interestingly, molluscs can grow quite large either with or without a shell. We've already noted the almost shell-less giant squid. There's also a giant clam, whose huge valves can be more than a metre across. The biggest molluscs on land are smaller than their marine counterparts, but still impressive. For example the giant African land snail can grow up to about 30 centimetres (a foot) in length. And there is a giant slug that reaches a similar size. However, we can't look at the largest land-based bivalve or cephalopod, because those two great molluscan groups are composed entirely of aquatic member-species. As far as we are aware, they have given rise to no land-based descendant lineages, even short-lived ones. This provides an interesting contrast with gastropods; and also with vertebrates, where an aquatic group – fish – gave rise to all the vertebrates on land, including us. We'll deal with vertebrate skeletons in the next section; but before we leave this one, let's consider one final point here that encompasses molluscs, vertebrates, and indeed animals in general. It's to do with the dimensions of 'size'.

Although there are some very large molluscs, there are no very tall ones. This is an interesting statement and leads us to enquire about the subjective nature of this thing we call height. The height of an animal is something that seems to apply more on the land than in the sea. We think of the height of a giraffe but the length of a whale. A giant squid is a very large creature, but we consider its size in terms of length, not height – even though it can stretch itself out fully in any direction, from horizontal to vertical, and can likewise swim in any direction.

Another aspect of the height-versus-length conundrum is the shape of the animal. We often talk about the length of a snake but rarely about its height – perhaps excepting poised cobras. So, it seems that when an animal is taller than long (giraffe) we use height but when the converse is true (snake) we use length. However, most land vertebrates have intermediate shapes where the two axes are of rather similar length. Horses are like this; typically, they're a bit longer than tall, yet there's a whole measurement system (hands) devoted to estimating their height.

The height–length issue is not just a philosophical or linguistic matter. It relates to the role of the skeleton, be it bones, armour, or shell. In one respect, two of these three types of hard part go together: molluscan shells and arthropod exoskeletons help protect against predation, while vertebrate bones do not (with some exceptions, such as in turtles). But in another respect, a different two go together: bones and armour enable land animals to get tall, while shells do not. It's interesting that the largest land snails and land slugs have similar lengths. While stretched out in full crawl, the snail is *taller* than the slug, but only because its guts are held high over its locomotory part, the foot.

This line of thought brings us back to gravity, which is where this chapter started. Both the vertebrate endoskeleton and the arthropod exoskeleton help their bearers to become large, despite the downward pull of gravity. But a molluscan shell doesn't really do the same thing. Instead, a giant land snail has to carry its heavy shell, rather than the shell in a sense carrying it.

Its shell is good to retreat into if a predator is threatening; but in terms of gravity it's a burden rather than a support.

## Building with Bones

We now focus on the third main type of hard skeleton – vertebrate bones. Like some exoskeletons and almost all shells, these are biomineralized, though in this case with calcium phosphate rather than calcium carbonate. And, as in those other biomineralized hard parts, the minerals are interwoven with organic macromolecules. The main group of these molecules in vertebrates is the collagens – proteins that characterize not just bone but cartilage, tendons, ligaments, and other sorts of connective tissue as well.

In fact, in both evolutionary and developmental terms, bone starts as cartilage, though there are some ifs and buts to this statement. The earliest vertebrates probably had cartilaginous skeletons that did not ossify. This condition is reflected in the skeletons of today's lampreys, and in cartilaginous fish such as sharks, rays, and chimaeras. A human embryo initially has a skeleton composed of cartilage; our bones begin to ossify at various stages of development, from late fetal to early adult.

The main qualification that should be made to the developmental cartilage-to-bone story is that some bone starts life not as cartilage but as skin. And just to add to the complexity, some bony structures, notably skulls, are made in both ways – some of the bones that fuse to become the human skull have dermal origins, others cartilaginous ones.

In the mature skeleton of a land vertebrate, be it a dinosaur, a giraffe, or a human, there are more than 200 bones. Some fuse directly with each other, as in the skull and the pelvis; others connect via cartilaginous joints, as in the arms and legs. Bones serve as points of attachment for muscles. The muscles are controlled by nerves, and in some cases by the brain. Reflex reactions are at one end of the spectrum, with no brain involvement; the actions of fingers on keyboards are at the other end, such as in my writing this sentence, where the brain is very much involved. Hence we can see a connection between the subject of this chapter – skeletons – and the subject of the next – intelligence.

The earliest vertebrates known are from the same geological period as the earliest trilobites and the earliest gastropods – the Cambrian – and indeed all three currently date to around 520 mya, though of course this figure may change if earlier fossils are found. The earliest vertebrates were jawless fish-like creatures. To get from that skeletal layout to ours there were several notable changes. First, there was the evolution of jaws, which occurred between 500 and 400 mya, giving rise to the group named after this evolutionary invention – the gnathostomes. Then there was the fin-to-limb transition that accompanied the vertebrate invasion of the land between 400 and 350 mya, hence producing the group called the tetrapods (Figure 7). The tetrapod design has been modified in many ways, including the evolution of arms (primates) or wings (birds, bats) from forelegs, the loss of limbs (snakes, many groups of lizards), and the conversion of limbs into flippers (penguins, dolphins, seals).

For those vertebrates that remained walking on land, or took to flying, the basic design of early tetrapods has proved very successful. The one–two pattern of long bones in the limbs (humerus followed by radius and ulna in your arm, for example) has provided a remarkably consistent basis for limbs of many kinds in amphibians, reptiles, birds, and mammals, albeit the number of digits has evolved many variations.

To get to a mammal from a reptile, the skeletal changes involved are relatively subtle compared to turning fins into legs – they could be described as quantitative rather than qualitative. They include: legs being pointed more downward than outward; a different pattern of holes in the skull through which muscles pass; and the evolution of a new jaw joint, with the bones of the old joint shrinking and being redirected into the ear. These changes, and others, took place in stages between about 300 and 200 mya.

The skeleton of the first primate was probably not very different from that of a mammal of the forest floor. But as primates

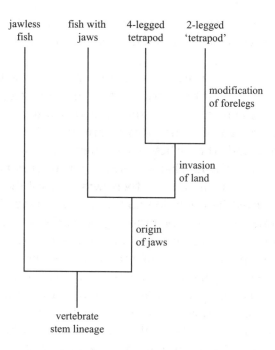

**Figure 7** Evolutionary modification of the vertebrate skeleton. An ancestral jawless vertebrate gave rise to a lineage with jaws. This lineage radiated into many forms, some (fish with jaws) remaining in the sea, others (tetrapods) colonizing the land. Many tetrapods retained the four-legged arrangement of the early amphibians, while others modified their forelegs for purposes other than walking and hence became two-legged 'tetrapods'. The top right branch of the diagram can be interpreted as showing the evolution of humans, birds, or any other two-legged group. The use of inverted commas shows the problem inherent in naming a taxonomic group after a body layout: not all tetrapods (in the taxonomic sense) have four legs (in the anatomical sense). Indeed some of them, such as snakes, have no legs at all.

took to the trees their skeletons became modified for their new-found arboreal existence. When proto-humans returned to the ground their skeleton was modified again, but interestingly not back to where it started before their primate ancestors became tree-climbers. Instead of returning to quadrupedal locomotion on the land we adopted its bipedal counterpart. Thus the hands that first evolved to grasp branches were able to make tools. It's an interesting question whether this habitat-related U-turn – up into the trees and back down again – was an essential evolutionary journey in the origin of human intelligence.

## Alternative Strategies

Let's end this chapter with a quick survey of some skeletal strategies that differ from the three we've looked at so far – vertebrate endoskeletons, arthropod exoskeletons, and molluscan shells. These include alternative types of hard part, and also a type of skeleton that paradoxically lacks hard parts altogether.

Coral reefs are produced by tiny colonial animals – the 'stony corals' – that are close relatives of animals entirely lacking in hard parts such as jellyfish. A reef is effectively a huge colonial skeleton. As in the skeletons we've already looked at, the colonial skeleton of a coral is biomineralized, and, as with shells and suits of armour, the main mineral is calcium carbonate. The individual coral creatures – polyps – that make up the colony are typically small, often just a few millimetres across.

Each polyp produces a mineralized exoskeletal compartment around itself. Although this is small, the skeleton of the whole colony is much larger; and the product of a group of neighbouring colonies is larger still – a reef. It's hard to believe that such huge structures are made by such tiny animals, but it's true. However, reefs eventually transcend their coral origin in that the hard material made by the coral polyps becomes colonized by all manner of other organisms, most of which are more conspicuous than the corals themselves. Sponges are among the non-coral sessile animals that grow directly on the reef; many of these have their own skeletons.

It's thought that the very first animal was a sort of protosponge. And today's sponges remain among the simplest animal forms on the planet in terms of their small number of cell types and their lack of organs. Nevertheless, most sponges have an internal assemblage of hard parts forming a sort of endoskeleton. The component parts are called spicules; they take a wide variety of shapes. A common one – the triaxon – is like a three-pointed star. But there are also simpler shapes, like rods, and more complex ones resembling hearts, spheres, and anchors. It is doubtful that sponges would have reached their current maximum size of more than 2 metres in height without their skeleton of spicules. Interestingly, while the spicules of some sponges are calcareous, others are siliceous – that is, they are made of silica ($SiO_2$) rather than calcium carbonate.

Now we come to the paradoxical case of the skeleton without hard parts. Although it sounds like a contradiction, it's not. Many groups of worms have what's called a hydrostatic skeleton. This phrase conveys the idea that a fluid-filled cavity has some of the properties of a skeleton; in particular, it can serve as a form of resistance to muscular contraction that enables movement. A typical worm has longitudinal and circular muscles. A pattern of alternate contraction of these, pressing in different ways on the fluid-filled body cavity, can make the worm move forward or backward. More complex writhing motions are also possible, and these are often assisted, as in the case of earthworms, by the body being divided up into segments.

Although a hydrostatic skeleton is of considerable use in movement, it's not much use in resisting gravity; as a result there are no tall animals built on this design. Then again, from a tree's perspective there are no tall animals at all – as we've seen, redwoods are many times the height of the largest ever dinosaur. Plants have reached heights that none of the other kingdoms of the living world can compete with, and there's a one-word reason for this: wood.

Wood is a familiar thing, but what exactly is it? We know it's hard, but what biochemical basis underlies its hardness? The answer to both of these questions is that wood is composed of a mixture of two main types of large molecule. One is cellulose, which is a carbohydrate composed of long chains of hundreds of glucose units. The other is lignin, an interesting substance because it does not belong to any of the four normally encountered classes of biological macromolecules – proteins, fats, carbohydrates, and nucleic acids. Instead, it is a cross-linked polymer of phenols. In terms of their quantitative contributions to wood, cellulose makes up about 50%, lignin about 25%. But it's the lignin that is the main contributor to gravity-resistance, as it is relatively incompressible.

What might the skeletal strategies be among life-forms on other planets? No doubt the answer to this question will vary from one life-bearing planet to another, and one reason for this variation will be different gravitational pulls, which in turn depend on different planetary masses. But while it's hard to predict either the nature or the strength of extraterrestrial skeletons, their existence most certainly can be predicted, simply because of the universality of gravity. Here's what I call the Ubiquitous Skeletons hypothesis: large organisms, whichever planet they are found on, possess skeletal structures, whether bone, wood, or 'other', which enable them to resist the downward force of gravity.

# 8 INTELLIGENCE AND LIFE

## What is Intelligence?

As in Chapter 1, where we arrived at a working definition of life, here we need to come up with a working definition of intelligence. The starting point for this is that we all have a general idea of what is meant by the term, and few people would dispute the direction of certain interspecies comparisons – for example humans being more intelligent than chimps, and chimps more intelligent than mice. From this starting point we can develop the idea further by listing some of the main features that are generally associated with intelligence: abstract thought, planning, reasoning, learning, memory, and self-awareness.

The problem is that we can't directly measure these things, which are all going on in the brain. Rather, we have to infer them from externally visible phenomena such as language and behaviour. Tool use is a behavioural activity that is helpful in this respect in comparisons of intelligence between different kinds of animal, as noted in Chapter 3. Making tools for later use implies planning ahead. Using tools in a gradually better way implies learning. And the more complex the tool, the higher the intelligence behind it is likely to be.

Importantly, 'intelligence' is a continuum rather than a binary variable. So we don't describe some animals as intelligent, others not. Rather, each species can be placed somewhere on the continuum. However, the situation is complicated by the existence of many animal species without brains (e.g. jellyfish) and even without nervous systems (e.g. sponges). Since a brain is a prerequisite for intelligence as normally defined in the living world, there are by definition many species that don't even make it onto the lowest point of the continuum. But we can ignore this complication here, as we're going to focus on animals that are near its upper end.

Another complication is that the different features that contribute to our broad notion of intelligence are themselves continua to various degrees; and they don't necessarily run in parallel. For example, ants have been observed using tools, including bits of leaf as carrying tools to get water or liquid food into their nests from the surrounding terrain. However, ants would probably score very low (zero?) on the continuum of animal self-awareness. So, ranking different animal species on an overall continuum of 'intelligence' may not always work.

Our exploration of animal intelligence here is a prelude to considering (later in the book) what we might expect to find on other planets in terms of intelligent life. And one interesting question in this respect is whether we might find life-forms more intelligent than humans. The answer almost certainly depends on the age of the planet. The older it is, the greater the potential for intelligent life to have evolved. Of course, we don't know to what extent an evolutionary process on another planet will be characterized by similar rates of change to the process here on Earth. Nevertheless, it seems reasonable to assume that intelligent life must always evolve from pre-existing unintelligent life.

We should acknowledge the possibility that an evolutionary process might run its course on a planet and not produce any significant degree of intelligence at all. If the Earth had ended its life (for whatever reason) at three billion years, that would have been the outcome here. In cases where highly intelligent life-forms do arise, they could potentially arise from different branches of the evolutionary tree concerned, and at different times. Let's now take a look at where the most significant animal intelligences on Earth have arisen, against the background of the evolutionary tree of life on Earth and, more specifically, the evolutionary tree of the animal kingdom.

## Surveying the Kingdom

Starting with the broadest groups into which the present-day biota can be divided – domains – we note that in only one of these – Eukarya – did intelligence arise. Looking at the various kingdoms of the Eukarya, including animals, plants, fungi, and others, we see that again only one – Animalia – is characterized by the inclusion of some intelligent creatures. But now if we focus on Animalia, and consider where among its major groups – phyla – intelligent forms have arisen, the answer is 'at least two': the chordates (in particular the vertebrate subphylum) and the molluscs (in particular the class Cephalopoda).

But should the answer in fact be 'more than two'? After all, we've already noted the use of simple tools by ants. Here we need to remind ourselves again of the fact that intelligence is a continuum. Thus the number of animal phyla in which 'intelligence' has arisen depends on where a line is drawn across this continuum, as we saw in Chapter 3. The drawing of a line isn't essential, but it's helpful if we want to focus – as we do here – on the top end of the spectrum. So, let's draw our line just beneath the octopus. In that case, 'at least two' becomes 'exactly two' rather than 'more than two' – because no other invertebrates can match the octopus in intelligence.

So, two phyla of animals include intelligent forms, defined in this way. How many phyla do not? Here we run up against the subjective nature of the taxonomic rank that we call the phylum. However, most accounts of the animal kingdom agree that there are about 35 groups of animals that are sufficiently distinct from each other to be considered as different phyla. There's a little variation among authors, of course, but the number is almost always between 30 and 40. Examples of animal phyla that are unfamiliar outside the world of professional biology include Bryozoa (moss animals), Rotifera (wheel animals), and Ctenophora (comb-jellies).

The evolutionary lineages that led to the groups we recognize as phyla diverged from each other early in animal evolution. Most were in existence 500 mya. Indeed, perhaps *all* were in existence then, though it's hard to be sure because many animal groups consist solely of soft-bodied forms and thus leave little trace in the fossil record. If indeed all 35 or so phyla were represented in the Cambrian fauna of 500 mya, they did not split 33/2, as they do today, into those with and without octopus-level intelligence. Rather, they split 35/0. This is because the cephalopod group that all species of octopus and squid belong to – the coleoids – had not yet arisen, and neither had the land vertebrates – the tetrapods. It seems a safe bet that the jawless fish of Cambrian oceans were less intelligent than an extant octopus – albeit we have no way to measure the intelligence of animals that are extinct.

Let's now concentrate on the two most interesting groups, the cephalopods and the vertebrates, in turn. The evolutionary tree of the cephalopods has several main branches. These include one to the nautiloids (with just one surviving genus, *Nautilus*), one to the extinct ammonoids, one to the ten-appendage group (squid and cuttlefish), and one to the eight-armed octopus group. This last group contains the most intelligent species of cephalopods, including the 100 or so species of the genus *Octopus*. Much of the work conducted on octopus intelligence has involved the species *Octopus vulgaris*. Given that 'octopus' can mean a member of this species, any of the 100 species of this genus, or even any of the 300 or so species belonging to the group Octopoda, we need to bear in mind possible differences in intelligence between species. After all, there are also about 300 species of primates, and we don't talk about 'primate intelligence' in a general way, as if lemurs and humans were similar.

The evolutionary tree of the vertebrates has branches leading to various groups, including some that are entirely aquatic (lampreys, cartilaginous fish) and some that are largely terrestrial (birds, mammals) but with aquatic subgroups (e.g. penguins, whales). The link between intelligence and habitat type is complex. Generally, the greatest intelligence is to be found in mammals and birds, groups whose origin (from different reptilian stem lineages) was on land, as are most of their extant members. Among the birds, the most intelligent forms are terrestrial – crows and parrots – though that said it's not an easy matter to compare the intelligence of a crow and a penguin. Among mammals, the two groups containing animals with greatest intelligence are the primates and the cetaceans (dolphins and whales), in other words one terrestrial and one aquatic.

This combined pattern of the emergence of intelligence across the vertebrates and the cephalopods defies any simple prescriptive rules. In other words, we can't really say, in mathematicians' lingo, that X, Y, and Z are the 'necessary and sufficient conditions' for high intelligence to evolve. Some things are necessary, most obviously a reasonably large brain. Other things may be helpful but unnecessary. For example, the evolution of manipulative hands with fingers (many primates) or arms with suckers (octopuses) may tend to be associated with the evolution of increased intelligence. But given the high intelligence of dolphins, they're clearly not essential.

Interestingly, intelligence – in the sense of the cephalopod level or beyond – has never arisen in a group of animals that lack eyes. While this cannot be interpreted as meaning that the processing of visual information is a prerequisite for high intelligence, it is certainly suggestive of a link (more on this in Chapter 20). And the similarity of vertebrate and cephalopod eyes is extraordinary, given that these evolved independently from each other. In both cases the result was a camera-type eye; a minor difference is that the cephalopod eye lacks a blind spot because the nerve fibres from the retina sensibly emerge at the back rather than at the front.

## Intelligent Molluscs

Having noted that intelligence varies somewhat across the few hundred species of octopus, even more so across the cephalopods, and hugely among the molluscs in general, we'll now focus on the peak of molluscan intelligence – that of the genus *Octopus* and its close relatives. Here, many observations of behaviour have shown play, learning, planning, tool use, and recognition of individuals. Some of these observations were conducted in the lab, others in the field.

One fascinating field study involved observations by divers, spread over a ten-year period, of the veined octopus (*Amphiocto-pus marginatus*) at two locations in coastal Indonesian waters – Sulawesi and Bali – at depths down to about 20 metres. At these locations, coconut-shell halves were available, as a result of human activity. Octopuses were observed using water jets to flush these out of the sandy substratum. Then they carried the half-shells from the location they found them in to another one, where they used them as a protective shelter. Sometimes, they carried half-shells singly, sometimes in pairs or groups of three. To carry the shells, they wrapped some of their arms around them and carried them underneath the body, using other arms to 'stilt-walk'. When the half-shells had reached their destination, the octopuses assembled them into pairs, inside which they hid.

The authors of this study – Julian Finn and his colleagues – interpreted these behaviours as demonstrating forward planning of tool use, which seems entirely reasonable. The initial stages of the endeavour – extracting the shell halves and carrying them – required expenditure of energy. The benefit only came later, with assembly of a shelter. Of course, octopuses would not usually encounter coconut shells in their habitat, but the authors suggest they may well exhibit the same behaviour with discarded large bivalve shells at other locations. If this is so, there's a strange irony: an animal that lost its shell – evolutionarily speaking – but gained intelligence uses its new-found intelligence to construct a shelter made with the shells of its dead molluscan cousins.

Various studies conducted in aquaria have focused on the ability of an octopus to open a screw-top jar with potential prey, such as a crab, inside. When first confronted with this challenge, octopuses often take several minutes to succeed. However, with practice they become quicker and learn to achieve the desired outcome – grabbing the crab – within seconds rather than minutes. One study even used a child-proof screw-cap – the type you have to squeeze at a particular point before a screwing motion will work. The Pacific octopuses involved managed to solve the problem in a little under an hour; and with practice they got the time down to about five minutes.

Another aspect of octopus learning has been demonstrated by maze experiments. One of these, conducted by Jean Boal and her colleagues in 2000, gave octopuses a choice of possible escape routes from an experimental arena, only one of which was open, the others blocked. The rationale behind the experiment was that the species concerned was a shy nocturnal one; the arena was brightly lit, while the escape route led to a small dark recess. Most of the octopuses learned to choose the open escape route over the course of several trials. However, interestingly, not all did. This result emphasizes the point that individual variation is an important factor in interpreting the results of experimental work on animal behaviour.

The same phenomenon – inter-individual variation – was observed in an investigation of 'play' in another species of octopus, carried out by Jennifer Mather and Roland Anderson. Here, some octopuses were seen to exhibit exploratory play with pill bottles that were available to them in an aquarium setting. They appeared to be playing with the bottles in much the same way as a child bounces a ball. This is significant because play may be linked to the development of consciousness. But are octopuses actually conscious? We can't really answer that question yet, and, as with intelligence, consciousness is a spectrum of possibilities, not a binary yes/no variable. Anyhow, my money is on these wonderful animals having by far the most advanced consciousness among all the million-plus species of invertebrates.

## Tools and Crows

Although octopuses are especially intelligent cephalopods, many other kinds of cephalopod come close – especially some squid and cuttlefish. Is the situation the same in birds? Cephalopoda and Aves are taxa of broadly equivalent rank – they are both conventionally thought of as classes (within the phyla Mollusca and Chordata respectively). So we can ask whether birds are also quite generally intelligent, or whether intelligence is more patchily distributed in this particular class of animals. The answer would appear to be 'a bit of both'. Inasmuch as there is patchiness, the crow family (Corvidae) stands out from most others.

This is not to say that tool use is rare among birds – it's not. Many years ago, as a PhD student studying populations of snails on sand dunes, I often came across what are known as anvil stones used by thrushes (family Turdidae). These birds smash the snail shells on the stone to break them and thus to get at the meaty interior more easily than is possible when the shell is intact and the snail can retreat into it. A similar kind of behaviour occurs in Egyptian vultures, which throw pebbles at large eggs – for example ostrich eggs – to smash their shells. Also, one species of Darwin's finches – the woodpecker finch – is famous for being able to use cactus spines or twigs to oust invertebrate prey items from crevices.

These examples all involve relatively simple behaviour, involving the use of a single tool to do a single job. In the crow family, more complex behaviour involving several tools has been observed. And the most impressive such tool use has been observed in artificial arenas in which the range of potential tools available would not have been encountered in the wild. In one study, by Lucas Bluff and his colleagues, New Caledonian crows were seen to engage in serial tool use – using one tool to obtain another, which in turn was used to obtain a third tool, which was used for a particular task. The task was to retrieve a food item from a box. The experimental setup was designed so that only a long stick with a hook at one end would work. But to get this type of tool the birds had to use a short tool to get a medium-length one out of another box and use the medium one to get the long tool out of its box. Having obtained the three tools in sequence, the crows then used the long hooked tool the right way round to get the food. This behaviour demonstrates the crows' ability to distinguish tools on the basis of size and shape.

No-one knows what's going on in a bird's mind. When a crow is engaging in serial tool use, what is it thinking? And is it self-aware? Another species of the crow family, the Eurasian magpie, is the only bird so far to have passed the 'mirror mark test' – all the other animals to have passed it are mammals (see next section).

The mirror mark test, developed by American psychologist Gordon Gallup, is as follows: an animal in an arena that contains a mirror has a conspicuous mark put on it – perhaps a sticky disc of brightly coloured paper or a chalk mark. If the animal starts trying to remove the mark, it is considered to have in some sense understood that the image in the mirror is itself and not another individual. Some magpies try to remove coloured marks with their beaks and, having failed, try to scratch the marks off with their claws. This is in marked contrast to the behaviour of other bird species, which generally show aggressive behaviour to the individual in the mirror, having assumed that it is another bird. Indeed, many of us have observed unintentional mirror self-recognition tests (albeit without marks) in our gardens. For example, it is not uncommon to see a blackbird repeatedly attacking an 'intruder' in its territory, which is merely its own reflection in a window.

The family Corvidae includes more than 100 species, spread across all continents except Antarctica. These include, in addition to crows and magpies, rooks, ravens, jackdaws, and jays. Most of these have not been subjected to the mirror mark test, but a few of them have, and have failed. At least one species of parrot and one of the tit family (Paridae) have also failed. So the Eurasian magpie is for now the only apparently self-recognizing bird. It seems reasonable to suspect that in the fullness of time at least a few others will pass the mirror test too. But for now, if we want to look at other self-recognizing species of animals, we need to turn to mammals.

## Mirrors and Mammals

There are about 5500 species of mammals. These are divided up rather unequally among about 20 orders (the level of taxon immediately above families). More than half of all mammalian species are rodents (2000+) or bats (1000+). There are a few hundred species of primates. Because the pattern of mammal diversification was rather bush-like, with many lineages diverging from each other within a relatively short amount of evolutionary time, mammal relationships have been notoriously difficult to determine, though much progress has been made in recent years through the use of DNA sequence data.

To see the nature of the problem, let's consider the three groups of mammals just mentioned – rodents, bats, and primates, and add another three – elephants, sea-cows, and dolphins, which belong in different orders (both from each other and from the three already mentioned). Of these six groups, two are aquatic. But habitat gives us no clues about relationships in this case. The two aquatic groups, dolphins and sea-cows, are not closely related to each other. Rather, each aquatic group is most closely related to a terrestrial one. If we bracket our six chosen groups into three pairs based on closeness of relationship, the pairs are: (primates + rodents), (dolphins + bats), (elephants + sea-cows), as shown in Figure 8. Of the many ways there are of choosing three pairs out of six groups, few people would pick this arrangement based on the general forms of the mammals concerned.

Passing the mirror test maps very unevenly to a mammalian evolutionary tree; but then so too does choice of species to put to the test. Far more primate species have been tried, for example, than species of rodents. In general, biologists choose those species that they hope will pass rather than those they suspect will fail. Nevertheless, some interesting facts emerge about mammals and mirrors, as follows.

Within primates, there is a clear difference between the great apes and all the others (including one gibbon and several monkey species that were tested). All great ape species pass the test, while all other primates fail it. However, as we've already noted, it's important to pay attention to variation between individuals within a species when interpreting the results of psychological testing on animals. Among the great apes, individual gorillas often fail the test. Some silverback males have been

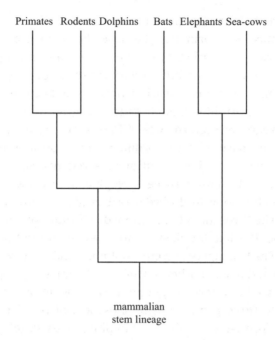

Primates  Rodents  Dolphins    Bats  Elephants  Sea-cows

mammalian
stem lineage

**Figure 8** Pattern of relatedness among six groups of extant mammals. This shows that habitat is no guide to closeness of relationship. The two aquatic groups shown (dolphins and sea-cows) are not closely related to each other, and represent independent evolutionary transitions towards an aquatic existence. With regard to self-awareness as revealed by reactions to mirrors, this has been observed in the left-hand group of each pair (primates, dolphins, elephants) but not in their right-hand counterparts (rodents, bats, sea-cows).

observed to exhibit solely aggressive behaviour, repeatedly charging at their perceived rivals, only to crash into the mirror. And young humans typically fail the test. It seems that we humans don't generally pass until we are about 18 months old, and the age threshold seems to be culture-dependent, with infants of some cultures not passing until a significantly greater age.

Outside the primates, the only mammals to have passed some version of the mirror test conclusively so far are bottlenose dolphins and Asian elephants. This means that all three pairs of orders we identified above have at least one species that passes. A reasonable conclusion from this pattern is that self-recognition has arisen at least three times in mammalian evolution.

The elephants subjected to the mirror mark test tried to remove the marks with their trunks. The dolphins lacked the ability to remove the marks on most of the body positions at which they were placed, so their passing the test was based not on removal but on investigation of the mark, as demonstrated by the American biologists Diana Reiss and Lori Marino, who quantified the mark-related behaviour of individual bottlenose dolphins. In addition to the quantitative data provided by these authors, qualitative information can be useful too, though we should always bear in mind its limitations. Anyone who has watched a video of a dolphin inspecting itself in a mirror will have little doubt that dolphins are self-aware creatures. Even without the diagnostic power of a mark, it's readily apparent from a whole range of behaviours (including inspection of the insides of their mouths) that a dolphin knows that the image it sees is one of itself.

## Space Travel and Apes

On 16 July 1969, a very large and complex tool was sitting on the ground at the Kennedy Space Center in Florida. This 'tool' included a Saturn V rocket and two modules. The purpose of the overall tool, of course, was to land the first humans on the Moon. And as we all know (except for the conspiracy-theorists), the compound tool achieved its aim brilliantly, despite having less computing power than today's average cell-phone.

It's quite a journey from using a stone hand-axe for hacking at carcasses to using a Saturn V rocket to boost people and machines off the Earth *en route* to the Moon. And yet both are tools – objects that are made and used to achieve a goal. Hand-axes were probably multifunctional tools; they seem to have been the first tools to have been widely manufactured by proto-humans, with their construction and use dating back to more than 1.5 mya. The complexity of tools has increased over the course of human evolution since those early days, but certainly not in a linear manner. The first wheel is thought to have been made only a few thousand years ago; the first telescope a few hundred years ago (around the year 1600); and the first computer just decades ago. We seem to be seeing an exponential increase in tool complexity over time.

But it's important to note the *range* of complexity of tools as well as the maximum complexity. As I write this, my coffee mug rests on a slab of Connemara marble, whose purpose is to stop my desk from blistering due to heat from the mug. This makeshift coaster has much in common with a hand-axe – it's a fashioned, flattish piece of stone. There are some papers on the desk held together with a paperclip – a small and simple tool that is still in widespread use – billions of them – around the world. Many of our tools are simple but effective. Then again, others are highly complex – and not just those of rocket science. Phones, computers, printers, lamps, maps, and dictionaries, to name but a few that I can see around me as I write this, are at various points on the ladder of complexity between stone coaster and spacecraft.

The first hominid launched into space by a rocket wasn't a human; it was a chimpanzee called Ham. His trip took place on 31 January 1961 (landing safely 16 hours after launch), only a couple of months before another hominid, Yuri Gagarin, completed an orbit of the Earth on 12 April of the same year. These days, the terms 'hominid' and 'great ape' are interchangeable, because they both refer to any descendant of the lineage that split off from gibbons about 20 mya. The family Hominidae thus includes humans and all species of chimp, gorilla, and orangutan (8 in total). The gibbon family, the Hylobatidae, includes all extant species of gibbon (18 in total).

The inclusion of chimps in the US space programme emphasizes not so much chimps' intelligence, great though that is against the backcloth of mammals in general, but rather the great gulf between their intelligence and ours. In this respect, apes and octopuses are rather different. Many species of octopus have been used in behavioural experiments, as we saw in a previous section. They all excel by general molluscan standards, but no one of them outdoes all the others and constitutes a statistical outlier. But this 'outdoing' is very much the case for humans among the apes.

If we poke further at the gulf between the advanced tools of rocket science and simple tools such as various types of hammering device, we see the need to portray it as something that's not quite as simple as a gulf between species – it's also a gulf within a species, namely ours. Rocket science can be traced back at least to the Second World War, where it was an altogether more sinister endeavour than it is (for the most part) today. It emerged from the industrial society that characterized the so-called First World. But there were then, and there still are now, some human societies that have little or no technology. For example, Brazil has many uncontacted tribes living in the Amazonian rainforest to whom our technology would seem very alien indeed.

Although members of these tribes would seem significantly less intelligent than us on the basis of a tool-making criterion, they are of the same intelligence as us on the basis of a brain-size criterion – both we and they have a cranial capacity of about 1350 cubic centimetres. This observation leads to an interesting question: why did humans evolve such a large brain size, one that is more than three times the size of the chimp's brain? Although it ultimately made rocket science possible, evolution does not think ahead – a point emphasized in the title of Richard Dawkins's book *The Blind Watchmaker*. The first humans with a brain the size of about 1350 cubic centimetres lived more than 100,000 years ago, pre-dating by a long period not only industry but also agriculture. What use was such a brain to early *Homo sapiens*? We don't yet have a clear answer to this question; various hypotheses exist on the relative importance of ecological and social factors, as noted in a recent *Nature* paper by Mauricio González-Forero and Andy Gardner of the University of St Andrews. However, the question itself is useful because it leads to a more general and thus arguably even more important one: what is the relationship between intelligence and fitness?

## Intelligence and Fitness

Evolution on Earth, and probably elsewhere, is driven by natural selection. By definition, natural selection is the spread of fitter variants at the expense of less fit ones. But what is fit depends on the environment. In some ecological contexts large body size is advantageous, so over time a lineage moves in that direction; but in other contexts small body size is advantageous, so a lineage moves in that direction instead. Albatrosses are not fitter than hummingbirds – that's a meaningless comparison. Both huge and tiny birds (and other animals) have been successful in their own ways.

The same is true of intelligence, even though at first sight there would seem to be no selective advantage whatsoever to becoming less intelligent over evolutionary time. And yet that is exactly what has happened in some animal groups. At the level of animal phyla, one of the groups that is mostly closely related to us is the Echinodermata – the phylum that includes sea urchins, sea stars, brittle stars, sea cucumbers, and sea lilies. Unlike the very distantly related jellyfish (phylum Cnidaria), echinoderms had bilaterally symmetrical ancestors that had head and tail ends, the former bearing a brain, albeit a small one.

From that starting point, the echinoderm lineage took a direction that ultimately led to the loss of both the brain and the head. If we look closely at a present-day echinoderm – for example a sea star – we see an animal that has radial symmetry (usually but not always pentaradial) and a diffuse nerve network that pervades the whole body, but without a single head-end concentration of nerve cells that could be called a brain (since there is no head). Consequently, sea stars and their kin cannot engage in complex behaviour in the way that, for example, octopuses can.

Echinoderms have been quite successful; there are more than 7000 species of them in today's biosphere, though none have colonized the land. They have a variety of lifestyles. The sea lilies, as their name suggests, are sessile plant-like creatures.

It's not so surprising that these have no brains – many sessile animals in other groups are brainless or nearly so. But sea stars are active predators: they survive by breaking into large prey such as mussels rather than by filtering small food particles from the surrounding water. Surely a brain would be a help rather than a hindrance to such a predator?

Actually it probably would, and this fact, together with the fact that no sea stars have brains, reinforces the 'blind watchmaker' way in which evolution works. The early echinoderm brain was not lost in a sea star. Rather, it was lost in a lineage whose exact nature is unknown but may well have had more similarity to a present-day sea lily than to a sea star or sea urchin. Whatever the exact structure of the early echinoderms that lost their brains, and whatever their ecological roles, we must assume that a brain (and hence intelligence) was unnecessary. Now, hundreds of millions of years later, it might well be advantageous for a sea-star lineage to evolve a brain, but if the relevant variation isn't there, then this evolutionary route is closed to them.

A brain being 'not advantageous' isn't the same as a brain being 'detrimental' in terms of fitness. So we might ask: why does a brain get lost, in evolutionary terms, if there is no natural selection against it? This is part of a much larger question about evolution, namely: why do *any* structures that have been gained during the long battle to survive subsequently get lost? There are several reasons for evolutionary loss, including the fact that any structure carries with it certain risks that make its possession disadvantageous on an overall basis if the structure no longer has an active role to play. The human appendix is a case in point. Without modern medicine, this vestigial structure might be lost altogether, because possession of it can lead to appendicitis, and thus potentially to peritonitis, which can be fatal at a pre-reproductive age.

Another reason for the loss of functionless structures is that there may be a trade-off between committing cells, materials, and energy towards them and towards some other structures that *do* have a function. This is thought to be the reason for the loss of eyes in blind cave fish. These animals have highly developed chemosensory organs in the head, which are a lot more use to them in the dark recesses of a cave than eyes would be.

Returning to intelligence, the fact that it and the physical basis for it – the brain – can be downplayed or even lost altogether in some lineages should temper our hopes for the discovery of extraterrestrial intelligence. There is nothing about an evolutionary process that makes intelligence inevitable: this can be called the Unnecessary Intelligence hypothesis. Even if it begins to evolve in some lineages, these may later backtrack and become what we might call 'stupider' than their ancestors. Natural selection is not on a long-term quest for the ultimate brainy animal; rather it is on a short-term quest for leaving more surviving offspring than your neighbours. The plant and fungal kingdoms have produced impressively complex life-forms without any brains at all. And it's always worth remembering that the vast majority of organisms on Earth are microbes.

*Part III*

# Planetary Systems and Life

## Key Hypotheses

### The Autospermia Hypothesis

This is a generalized version of the Terraspermia hypothesis for all inhabited planets. Wherever life has evolved, its origin was on the same planet as its evolution. If successful space travel and planetary invasion by dormant spores occurs anywhere, it is only within very small-scale planetary systems, and even this seems rather unlikely.

### The Bio-Region Hypothesis

On a tidally locked planet, under certain atmospheric conditions, life may be able to originate, evolve, and prosper in a restricted region that is short of being a complete biosphere (or bio-hollow-shell).

### The Resilience of Life Hypothesis

Despite much discussion of ecological fragility in the face of human onslaught, many life-forms on Earth are very resilient, as testified by organisms that have evolved to live on toxic wastes and in other inhospitable places. Extraterrestrial life-forms are likely to be just as resilient as their counterparts on Earth.

# Part II

# Planetary Systems and Life

## Key Hypotheses

# 9 TYPES OF PLANETARY SYSTEM

## Our Local Example

Our own solar system is the natural starting point for looking at planetary systems in general. Although discoveries about our system are still being made, especially in its outer reaches, its main constituent bodies (in terms of mass) were known a long time ago. The five 'naked-eye planets' – Mercury, Venus, Mars, Jupiter, and Saturn – were known to the ancient Greeks, one of whom (there is debate on this, but it might have been Pythagoras) realized that the 'morning star' and the 'evening star' were the same body – the planet Venus. Uranus was discovered in 1781 by the Hanoverian-British astronomer William Herschel, and Neptune in 1846 by German astronomer Johann Galle, on the basis of a mathematical prediction of the planet sent to him by French astronomer Urbain le Verrier.

The Sun makes up more than 99% of the mass of the solar system. And the planets make up more than 99% of the rest of its mass. So everything else put together represents only a tiny fraction of the total. However, in terms of the number of objects this 'everything else' component is huge. There are at least five dwarf planets, including Ceres in the asteroid belt and Pluto in the Kuiper belt, with more probably still to be discovered. And there are more than 170 moons, most of them orbiting Jupiter or Saturn. The numbers of asteroids and comets are huge and hard to pin down. More than half a million such bodies are known. The exact number we recognize depends heavily on the lower size limit we use.

Before moving on to quantifying the spatial arrangement of the objects of the solar system, it's worth noting how planets are defined, given the demotion of Pluto in 2006. These days, a planet is defined by the International Astronomical Union as a body that (1) directly orbits the Sun (or another star in the case of exoplanets), (2) is massive enough to be gravitationally rounded rather than irregularly shaped, and (3) has cleared its neighbourhood of other bodies that directly orbit the Sun. Jupiter's four Galilean moons fail on the first count because they orbit Jupiter, while Pluto fails on the third, as do other dwarf planets. Most asteroids fail on both the second and the third – almost all of the bodies in the asteroid belt are irregular in shape.

We all have a rough idea of the layout of the solar system, in particular the order of the planets in terms of their distances out from the Sun. But as in our discussion of the Milky Way in Chapter 2, it's better to have some degree of quantification rather than just a rough idea. For the Milky Way we used the light year as a unit of distance. This unit (along with another called the parsec) is appropriate at galactic scales. But it's too large to use for planetary systems, because these systems – including our own – are typically much less than 1 light year across. However, miles and kilometres are too small. These units of distance were devised for use on the surface of a single planet; we would have to use many millions of them for measuring distances from planet to planet within the solar system.

The unit of distance that's most useful for measurements within a planetary system is the astronomical unit, or AU. This is the distance from the Earth to the Sun. Thus 1 AU is equal to approximately 150 million kilometres or 93 million miles. Of course, the distance from the Earth to the Sun is not fixed, because the Earth, like all the other planets, has an elliptical orbit. But we can ignore this and other complications for our purposes here. The important thing is that the AU is many orders of magnitude longer than a mile or kilometre; likewise, it is many orders of magnitude shorter than a light year (1 light year is about 63,000 AU).

Armed with this intermediate unit of distance, here are the approximate locations out from the Sun of the four rocky planets: Mercury < 0.5, Venus 0.7, Earth 1 (naturally), and Mars 1.5. The asteroid belt extends roughly between 2 and 4 AU, but with fuzzy edges. The gas giants (Jupiter and Saturn) are at approximately 5 and 10 AU, while the ice giants (Uranus and Neptune) are at about 20 and 30 AU. The Kuiper belt, in which Pluto and many other bodies orbit, extends from about 30 to 50 AU. Beyond this there are many 'scattered disc objects', whose elliptical orbits can extend out to very approximately 100 AU. We can take this measure as the edge of the solar system from our perspective of the search for life, though we need to recognize that many other choices of its edge are possible.

All the solar system's planets, dwarf planets, and asteroids orbit the Sun in the same direction – anticlockwise if viewed from 'above', in the sense of terrestrial north. This is a consequence of their common origin via accretion of smaller particles in the proto-planetary disc, which, 4.6 billion years ago, was rotating in this direction. (An interesting exception was recently discovered near Jupiter; this asteroid is thought to have come from another system a very long time ago.) Orbital speed slows down as we go outward, so in terms of the planets Mercury's speed is the highest, Neptune's the lowest. The reason why Neptune's year is 165 Earth years while Mercury's year is less than three Earth months is partly their differences in speed and partly their different lengths of orbit.

With regard to possible life, the vast majority of the solar system, like the vast majority of the galaxy, is of little interest to us. For the most part, our system looks barren. And we've found it to be barren when landing spacecraft in various places, including planets (e.g. multiple Mars landers), moons (ours, and Saturn's moon Titan), a comet (the European Space Agency's *Rosetta–Philae* mission to comet 67P, landed in 2014), and a few asteroids (the Japanese *Hyabusa* missions to the asteroids Itokawa and Ryugu, landed in 2005 and 2018 respectively; and the NASA mission to Bennu, arrived December 2018). But, in addition to the Earth, there may be a few other exceptions to this rule of general barrenness in our system. We'll have a closer look at these in the next chapter.

## The Central Sun

Other planetary systems may differ from ours in a number of ways. One of the most important of these is the nature of the central sun. Our own Sun is a quasi-average star in terms of mass. As we saw in Chapter 2, there is a skewed distribution of star masses, with very large stars 50 times the mass of the Sun being rarest, red dwarfs with a fraction of the mass of the Sun commonest. To look at the full range of stars that planets can orbit, we need to fill in the gaps between these two extremes; and we need to look at luminosity and temperature as well as mass.

There are several systems for classifying stars. A commonly used one is the Harvard system, devised by American astronomer Annie Jump Cannon at the start of the twentieth century (Figure 9). This is based on observing the light from a star; in particular, it is based on examining the relative strengths of spectral lines in the light. These strengths are determined by the temperature of the 'surface' of the star – the photosphere (which, like the biosphere, is not spherical, but rather takes the form of a shell). The Harvard system has seven main classes – O, B, A, F, G, K, and M – for which a useful mnemonic is 'Oh Be A Fine Guy/Girl, Kiss Me'. Stars in class O are the hottest, while those in class M (including red dwarfs) are the coolest. The surface temperatures are about 40,000 for class O, a mere 3000 for class M. The Sun, in class G, has a surface temperature of between 5500 and 6000. (All these temperatures are given on the Kelvin scale, which starts at absolute zero, or –273 degrees Celsius.) Because a number of features of stars co-vary, the Harvard system is useful for giving us an estimate not just of the temperature, but also of the mass, luminosity, and lifespan. These are also given in Figure 9, along with the approximate fraction of all stars that belong to each class.

From the perspective of an orbiting planet, there are two main effects of the nature of the central star. First, and rather

| Class | Colour | Temperature (K) | Mass (Suns) | Luminosity (Suns) | Lifespan (yrs) | Fraction |
|---|---|---|---|---|---|---|
| **O** | blue | >30,000 | >16 | >30,000 | <0.01 billion | <0.001% |
| **B** | blue-white | 10,000 – 30,000 | 2 – 16 | 100 – 30,000 | 0.1 billion | 0.1% |
| **A** | white | 7,500 – 10,000 | 1.4 – 2 | 10 – 100 | ~1 billion | 0.6% |
| **F** | yellow-white | 6,000 – 7,500 | 1.04 – 1.4 | 2 – 5 | ~5 billion | 3% |
| **G** | yellow | 5,200 – 6,000 | 0.8 – 1.04 | ~1 | ~10 billion | 8% |
| **K** | orange | 3,700 – 5,200 | 0.45 – 0.8 | <0.5 | ~100 billion | 12% |
| **M** | orange-red | 2,000 – 3,700 | 0.08 – 0.45 | <0.1 | trillions | 76% |

**Figure 9** The Harvard spectral classification of stars, showing various features of the seven main classes, together with the approximate fraction of all stars that belong to each class. Colour is a rough guide only. Mass and luminosity are given as multiples/fractions of the Sun's values. As can be seen, lifespans vary enormously, and this will have a major effect on the likelihood of life having time to evolve on orbiting planets.

obviously, at any given distance out from a star a planet will receive more heat and light the higher the spectral class into which the star falls. Second, and not obvious at all before we look at the penultimate column of Figure 9, the shorter will be the planet's life. This is a direct consequence of the short life of the star itself. Basically, big stars live fast and die young. Or, to put it more scientifically: the larger the mass of a star, the shorter its period of existence between its birth as a proto-star and its death throes as a giant or supergiant.

This relationship seems counterintuitive. The more massive a star, the greater its initial quantity of fuel – hydrogen – for nuclear fusion. So, other things being equal, we would expect a big star to live longer. However, other things are not equal. In particular, the rate of 'burning' the fuel varies. The most massive stars are hottest and have the highest burn-rates; and this outweighs the fact that they have more fuel in the first place. This effect is so pronounced that O-class stars only live for a few million years (or even less for the very biggest of them), whereas the Sun's lifespan will be about 10 billion, and that of a red dwarf hundreds of billions or even trillions. We'll return to this point in the next chapter, in relation to planetary habitability.

Before we get into the complications posed by having more than one star in a system (next section), there are a few points that need to be added. First, there's a difference between a star's luminosity and its perceived brightness, as we saw in Chapter 2: the former is an intrinsic feature of the star, the latter depends on its distance from the observer. Second, there's a difference between mass and size. When a star like our Sun swells up to become a red giant near the end of its life, it increases in radius (and hence size, in the sense of volume) but not in mass. Third, we've been talking, up to now, about the main-sequence life of stars – the phase in which they spend about 90% of their time (recall the Hertzsprung–Russell diagram, Figure 3). This is the only phase relevant to life on an orbiting planet; neither proto-stars nor the various forms of dying stars are likely to present a habitable zone for life-forms. Fourth, calling a star a dwarf can be simply a comment on its size, but it can be a comment on its life-history stage too. A red dwarf is a small main-sequence star. However, a white dwarf is the terminal stage of a main-sequence star like our Sun. It's also the terminal stage of a red dwarf star, so red dwarfs end up as white dwarfs. And, as we saw in Chapter 3, a brown dwarf is a sort of failed star – one whose generating cloud fragment was so small that on collapsing it never quite reached the threshold temperature for nuclear fusion to begin.

Finally, note that Figure 9 quantifies the skewed distribution of star masses. O-, B-, and A-class stars together make up less than 1% of the total. F-class stars represent about 3%. G-stars (including our Sun) are about 8%, and the slightly smaller and cooler K-stars are 12%. Together all these classes make up about a quarter of the total. So, about three-quarters of all main-sequence stars belong to class M, the class that includes red dwarfs. Given this, it's hardly surprising that the nearest star to the Sun, Proxima Centauri, is a red dwarf; and it's one that is now known to have an orbiting exoplanet – we will look at this planet in detail in Chapter 14.

## Binary Systems

A planetary system needn't have just a single central sun. Indeed, we now think that about half of the stars in our local area of the galaxy – and probably also in the galaxy (and universe) as a whole – occur as binaries or even as multiple systems, with the latter having at least three stars. Many of the stars we can see in the night sky turn out to be double or multiple when looked at through a telescope. For example, the brightest star of all, Sirius, is a binary. However, we need to be careful to distinguish between 'optical doubles' and true binary systems.

In a true binary, the stars are close enough together to be locked in each other's gravitational embrace. Both orbit the system's centre of gravity, which will be halfway between the two stars if they're of equal mass, or asymmetrically placed if they're not. In an optical double, the two stars may be a vast distance apart. This phenomenon arises when, in a particular direction from Earth, there is one star that is many light years away – say 20 – and another that is many more light years – say another 30 – beyond the first one. In this case, our Sun and the nearer star of the optical double are closer together than are the two stars of our perceived 'double'. So, optical doubles – like constellations – are accidental effects of the position of the observer, whereas binary systems are very real.

How far apart are the two stars of a binary, or the 3+ stars of a multiple system? To answer this question we use astronomical units again, not light years. In the case of Sirius, the separation of the two stars is about 20 AU, and their orbital period is about 50 Earth years. This is a very asymmetric system: it consists of a main-sequence A-class star and a white dwarf. Other binaries are less asymmetric. For example, what appears to us as the 'star' chi-Draco (in the constellation of Draco the dragon, close to the Pole Star) is in fact a binary of an F-class star that's just a little bigger than our Sun and a K-class star that's a little smaller. Here the separation is only about 1 AU, and the orbital period is about 9 months. The closest star system to us, Alpha Centauri, is a triple consisting of stars A and B, which are quite close together (the distance varies over time between about 10 and 40 AU), and a third star, Proxima Centauri, which is about 13,000 AU from the other two.

Since binary and multiple star systems are common, it seems reasonable to expect that a significant fraction of all planets is to be found in such systems, albeit we cannot yet quantify 'significant'. But how do such planets orbit? Even in a system of just two stars, there are three possibilities: orbiting star 1, orbiting star 2, or orbiting both, with the last of these being referred to as a circumbinary orbit. For three stars, there are up to seven types of planetary orbit in theory, though in practice there are often fewer than seven because of the relative distances apart of the stars. For example, in the Alpha Centauri system, the most likely planetary orbits would be of star A, star B, the AB pair of stars, or the much further-flung star C (Proxima Centauri). The planet Proxima b does the last of these things. Although no planet has yet been found orbiting A+B but not C, such an orbit seems entirely possible; while an orbit of A+C but not B doesn't seem possible at all.

Of the many multiple star systems discovered so far, in some cases it's not yet clear exactly how many stars are involved, while in others the stellar components of the system are well established. Some binary and multiple systems are already known to have planets; and many others will doubtless turn out to have planets following further study. The concept of a planetary habitable zone is necessarily more complex in a binary or multiple system than in one with a single central star. We'll look at habitable zones for planets in the next chapter; but before that we must look at planets themselves.

## Planetary Possibilities

The main message to have come out of the last three decades of exoplanet discovery is that anything is possible – well, almost.

The number of planets in a system varies; there is no 'norm'. The arrangements of the planets with respect to each other and to their local sun(s) is variable too, as we saw in Chapter 3. For example, gas giants can be far out, as in our own system, or close in, as in many others. Planet size is very varied, and will probably in the end turn out to be even more so than we can see at present, given the bias in our detection methods towards finding large planets. Orbital periods vary immensely. And the shape of the orbit varies, with some planets having almost circular orbits, others much more elliptical, or 'eccentric', ones. Given this enormous variation, a good way to characterize it without drowning in detail is to look at extremes. We can then reasonably assume that anything between the extremes occurs somewhere. And with future discoveries the current extremes themselves will be outdone.

Even just within our own system the variation among planets is considerable. In terms of volume, more than 1000 Earths could be fitted into Jupiter. The difference in masses is smaller (because Jupiter is gaseous) but still impressive – Jupiter's mass is equivalent to about 300 Earths. Neptune has an orbital radius 30 times that of our own. Maximum surface temperatures vary by hundreds of degrees – whichever scale we use. The Earth has a near-circular orbit, Mercury a much more elliptical one. Orbital period is longest for Neptune (165 years), shortest for Mercury (88 days). Jupiter has about 70 moons, Venus none. Of the four terrestrial planets, Earth and Venus have thick atmospheres, Mercury and Mars very thin ones – indeed Mercury's is so thin as to be virtually non-existent. The Earth has a strong magnetic field, while Mars does not. And so on.

Beyond our own system, the variation is even greater, as might be expected given the much larger number of planets involved. The most massive exoplanet discovered to date has a mass many times that of Jupiter; the least massive has a mass less than 10% that of the Earth. Orbital periods vary from a few hours to almost a million years. The least eccentric exoplanet orbit is, like that of Earth, close to being circular. The most so has an eccentricity of 0.97, similar to that of Halley's Comet, and way more elliptical than the most extreme local planet in this respect, Mercury, whose eccentricity is 0.2. (Orbital eccentricity varies on a scale from 0, for a circle, to 0.99, for a very elongated ellipse; a value of 1 or more indicates a parabola or hyperbola, which are paths of no return, as in the case of our recent interstellar visitor 'Oumuamua, which has an eccentricity of 1.2.)

Given the wide range of possible planets, it is clear that many of them are completely uninhabitable by life as we know it and indeed by any other kinds of life that might reasonably be imagined. But some – a small fraction of a very large number – are almost certainly life-friendly. However, life might not need a planet at all. Other bodies within planetary systems might also be habitable, again with the proviso of 'in just a small fraction of cases'.

## Other Orbiters

As we've already seen, the types of orbiter in our own system include planets, dwarf planets, moons, asteroids, and comets. All of these sorts of objects probably exist in other systems too, though our knowledge of most of them so far is poor. We know of a handful of exocomets; and there are some candidate (but as-yet unconfirmed) exomoons. Given the small size of most asteroids, and the very eccentric orbits of comets, exomoons are perhaps the most interesting from a looking-for-life perspective. However, most sizeable moons in our system, and probably in others, exhibit a phenomenon called tidal locking.

Although the phrase may not be familiar, the closest example of it is very much so. The Moon is tidally locked to the Earth. This is simply another way of stating that we always see one side of the Moon, never the other. Of course, the Moon is a quasi-sphere, not a disc, so the idea of 'sides' is perhaps a bit odd. Nevertheless, in terms of the Moon's overall surface area, we only ever see a little more than half of it. The reason for this is that over time the gravitational pull of the Earth has slowed the Moon's spin rate down to the point where a complete rotation on its axis takes the same length of time as a complete orbit of the Earth. If, instead of this, the situation was that the Moon had stopped spinning altogether, then we would in fact see almost all of its surface over the course of a month.

How does this apparent magic work? How does the Earth manage to slow the Moon's spin rate down, but not to zero? The answer is that the gravitational pull of the Earth distorts the Moon's shape so that its diameter from front to back (from the Earth's perspective) is slightly greater than its diameter from side to side. The bulge facing the Earth is initially moving as the Moon spins. But this in turn means that the bulge is always slightly misplaced, and acts as a sort of handle for the gravity of the Earth to tug on. The end result of this tugging process is synchronicity between the rotational and orbital periods of the Moon. This locking phenomenon is quite widespread. Tidal locking is also exhibited by other large moons in our system, including the four Galilean moons of Jupiter – Io, Europa, Ganymede, and Calisto – and the large moons of Saturn, including the biggest one of all, Titan, and the possibly life-bearing one, Enceladus. It is to be expected that many moons in other systems will also be tidally locked to their host planets.

Tidal locking can also occur between a planet and its sun, so that one face of the planet receives a huge amount of heat and light, the other almost none. There are no instances of this complete locking in our own system (although Mercury exhibits a related phenomenon called spin–orbit resonance, as we'll see later), but there are already known cases of it elsewhere. This situation is clearly not conducive to the existence of a biosphere like our own on the planet concerned. However, this does not necessarily mean that there can be no life, as we will see in the Bio-Region hypothesis of Chapter 11.

## On Being Alone

At several levels of spatial scale, astronomical entities tend to be found in groups, but there are always exceptions. Most galaxies group together in clusters and superclusters, but there are also 'field galaxies', each of which exists on its own – a field galaxy is not gravitationally bound to any others. Most stars are found within a galaxy, but again there are exceptions. Intergalactic stars are known from several locations, including in between the Milky Way and our nearest galactic neighbour, Andromeda.

There are different theories about how intergalactic stars arise. One plausible idea is that they often arise from interactions and mergers of galaxies. For example, about 25 million light years away from us is a pair of interacting galaxies, the larger one – the Whirlpool galaxy – a spiral, the smaller one a dwarf elliptical. Looking at this pair of galaxies through a telescope, whirling milky trails can be seen extending away into intergalactic space. These trails consist of many stars that have been extracted from their original home in one of the galaxies and thrown beyond both of them by changing gravitational forces.

As well as field galaxies and intergalactic stars, there are also lone planets – possibly rather a lot of them. These have been variously named orphan, rogue, or sunless planets. They may arise in two ways. First, they may be planets that were originally within a planetary system but were expelled at some point in their history. Second, they may have formed *in situ* by collapse of a small gas cloud – whether such an entity should be called a 'brown dwarf sun' or a 'big brown planet' is a moot point.

In terms of effects on potential habitability, the last of these aspects of being alone is probably the most important. It seems unlikely that sunless planets could provide the basis for a biosphere. Some of them are gaseous. Those that are solid have extremely low surface temperatures. And the light that provides the lion's share of the energy that organisms fix into biologically useful form on Earth is altogether absent. Could a sunless planet be host to a simple microbial biosphere based on chemosynthesis? That depends on whether there is some source of internal heat equivalent to that characterizing Earth's deep-sea hydrothermal-vent ecosystems, which derives from plate tectonics. Such life may be possible, and it has been hypothesized by a few authors, notably the New Zealand planetary scientist David Stevenson, and the American authors of the 2017 book *Exoplanets*, Michael Summers and James Trefil; however, as all these scientists point out, rogue-planet biospheres would be very hard to detect.

The effects on planetary habitability of orbiting an intergalactic star, or orbiting a star in a field galaxy, are probably minor. The most important body for a would-be habitable planet is its sun. Other important bodies will be its moon(s) and its fellow planets in the system concerned. For example, a large planet like Jupiter may be able to act as a gravitational magnet for wandering asteroids in its system and hence reduce the rate of their impacts on neighbouring smaller planets. In contrast, what lies beyond the system is in a sense academic.

# 10 HABITABLE ZONES

## Liquid and Life

Most matter that we encounter personally is in one of the three familiar states of solid, liquid, and gas. There's another state that's less familiar to us and yet very common in the galaxy and the wider universe – plasma. This is an ionized state that is found in the hot interiors of the Sun and other stars. All four of these states of matter are important for life. Without solar plasma, there would be no photosynthesis by plants. Without gas, land animals would be starved of oxygen, and land plants of carbon dioxide. We are all ultimately supported by a solid substratum, even if this support is indirect – via the sea-floor – for marine organisms. And liquid is essential for the biochemical reactions of metabolism.

What state of matter characterizes life-forms themselves? Organisms are probably best described as mixtures of liquids and solids. In large organisms the solid material is obvious – for example the various forms of support structures that we discussed in Chapter 7, including wood and bone. In smaller organisms the solids are less obvious but they are there nevertheless. The cell membrane of a bacterium, and the cell wall outside it, are solid rather than liquid. A worm has solid tissue such as muscle as well as liquid tissue such as blood. Within an animal or plant cell, the liquid cytosol is the main component of the cytoplasm, but suspended within it are organelles such as mitochondria, bounded by solid membranes.

In a gas giant planet, the problems for life – in terms of the available states of matter – are twofold. First, there is no solid surface to underpin everything. There may be a tiny planetary core, but if so its surface is too deeply buried in gases to be of use for supporting a biosphere. Second, there is no liquid, at least on whatever we call the surface, though there may be a bizarre 'metallic liquid' form of hydrogen further down. On a rocky planet there is no problem of a missing substratum, but usually there is no liquid and sometimes no gas – or negligible amounts of it. Mars provides a great solid surface for robotic vehicles like the *Curiosity* rover, and one day probably for humans too. However, present-day Mars lacks surface liquid, and the atmosphere is far too thin for any terrestrial organism to breathe – even if it had the right composition, which it doesn't.

So, what we really want for a potentially habitable planet is a three-way combination of solid, liquid, and gas. Usually, the emphasis is put on the necessity of liquid – and water in particular – because the other two are more often present. Venus and Mars both have solids and gases. But neither of them has any surface liquids, at least as far as we know. The surface of Venus is hidden by its dense atmosphere of carbon dioxide, but pictures taken by probes such as the Soviet *Venera 9* show a barren rocky landscape – hardly surprising since the average surface temperature is about 450 degrees Celsius. The only bodies in the solar system known to have liquid on the surface are the Earth (naturally) and Saturn's moon Titan. But in the latter case the liquid is not water. Rather, it's a mixture of organic compounds, probably dominated by methane, with lesser amounts of ethane and other hydrocarbons.

So, what we'd like to find on an exoplanet, if our interest lies in its potential habitability, is a solid surface, part of which is covered by liquid, surrounded by a reasonably dense atmosphere. For life as we know it, and perhaps for other forms of life too, the liquid has to be water. Our bodies are more than 50% water, as are the bodies of other terrestrial creatures. The various molecules of which we're made all interact with water in appropriate ways – for example by being soluble in it – so that the frenetic biochemical activity that we call metabolism can take place. Water has many properties that make it unique as a medium for life – as explained in a wonderful (and wonderfully short!) book by the British physicist John Finney. Perhaps there is some life somewhere that is based on some other solvent than water. As a potential example of this within our own solar system, perhaps methane-based life will one day be found on Titan; but there's no sign of it yet, and I must admit to being sceptical about its likelihood.

Another way to approach the 'liquid and life' connection is to look at what happens to Earthly organisms when liquid water is scarce or absent. But we've already done that in Chapter 1. As we saw there, a tardigrade can survive a total lack of water for an extended period, but to do so it has to go into a state of suspended animation (cryptobiosis) in which its metabolism is shut down. Clearly, reproductive activity is shut down too. In other words, the main defining processes of life – metabolism and reproduction – cannot be sustained (at least on Earth) without water. In most organisms, the result is death. In tardigrades and a few others, death can be postponed.

## Defining the Zone

Because of the dependence of Earth life – and perhaps all life – on water, the habitable zone around a star is defined by potential occurrence of liquid water on the surface of an orbiting planet (or other suitable rocky body, such as a large moon). This zone takes the form of a ring of space which, in the case of our own system, includes the orbit of the Earth. Clearly, if we define the zone in this way, the innermost and outermost parts of an extensive planetary system will be outside it – albeit the phrase 'outside the habitable zone' is ambiguous. In fact, 'inside the habitable zone' is ambiguous too. So let's adopt a usage that goes as follows. 'Inside the habitable zone' means *within* it, as in the case of the Earth. 'Outside the zone' means either external to it (like Jupiter) or internal to it (like Mercury).

The habitable zone has several other names. It's sometimes called the Goldilocks zone, the link being the porridge or planet that is not too hot or too cold, but 'just right'. It's also sometimes called the 'potentially habitable zone', which is technically correct but a bit of a mouthful. 'Ecosphere' and 'liquid water zone' have also been used, but the former is likely to be confused with other usages (e.g. experimental ecological microcosms) and the latter doesn't quite work without some restriction to planetary *surfaces*. Finally, the habitable zone has sometimes been called the 'temperate zone' – for example in the 2017 book *The Planet Factory* by astrophysicist Elizabeth Tasker. The rationale for this last usage is to emphasize the 'potential', and to avoid leaping to the conclusion that planets in a habitable zone are necessarily inhabited. While this is a commendable rationale, in my view the term 'temperate zone' causes more problems than it cures. In particular, it doesn't connect well with traditional usage of 'temperate zone' for the Earth. All of the tropical, temperate, and polar zones of the Earth are habitable; and in terms of biodiversity, it's the tropical zone, not the temperate zone, that wins. So herein I will stick to the most-used and in my view best term, the habitable zone – which we owe to the Chinese-American astrophysicist Su-Shu Huang, who first used it in 1959.

The main factor influencing whether water is in solid, liquid, or gaseous form on a planet's surface is temperature. It's not the only factor though. Recall that at the bottom of Earth's deep oceans liquid water is often found at temperatures above 100 degrees Celsius. The reason it stays liquid and doesn't boil is the extremely high pressure. However, on the surfaces of Earth-sized rocky planets, the pressure will normally not be so extreme. So we focus mainly on temperature in working out where we think the habitable zone lies around any particular star.

Clearly, the habitable zone will be closer in for less luminous stars; and further out for more luminous ones. The best unit to use for estimates of the inner and outer boundaries of any particular habitable zone is the astronomical unit (AU), which was introduced in the previous chapter. We'd do well to remember that in all cases, even in our own system, these are indeed *estimates* (based on modelling), not precise measurements, and they generally have quite wide errors associated with them.

Many diagrams of our own habitable zone show Earth as the only planet inside it. However, some show Venus as being just inside its inner margin and/or Mars as being just inside its outer margin. Depicting Venus within the zone is overly optimistic, but doing so for Mars is not. The Sun's habitable zone extends at least from 0.95 to 1.65 AU, and it may, depending on various factors, extend further – from about 0.8 to 2 AU, as noted by the American planetary scientist James Kasting, in a research paper of 1993 and in Chapter 10 of his book *How to Find a Habitable Planet*. Even the more conservative of these ranges includes most of the orbit of Mars (an ellipse with distance from the Sun to the planet varying from 1.38 to 1.67 AU).

For planetary systems orbiting a red dwarf, the habitable zone (all of it, including its outermost domain) will typically be much less than 1 AU from the star. At the other extreme, planets orbiting large, luminous stars of class O would need to be over 100 AU out to be habitable. And there's a continuum from one of these extremes to the other. Having said that, the situation is not as simple as the habitable zone simply shifting out with increasing stellar mass. There may be problems with any planets at all being habitable if they orbit stars that are either very large or very small. We'll deal with those problems shortly.

Habitable zones are normally depicted as being two-dimensional equivalents of a biosphere. That is, they are round, flat, and have a hole in the middle. When the star being orbited is on its own, that's probably about right, though we perhaps shouldn't rule out the possibility that a few systems might have their constituent planets orbiting at considerably more inclined angles than our own, in which case the flatness may not be appropriate. In these cases the habitable zone should be depicted as having some depth. In the extreme case of *very* inclined orbits, we would have a habitable sphere, or, more correctly, a habitable spherical shell – though this is unlikely to apply often, given the starting point of a planetary system as a proto-planetary disc.

In the case of binary and multiple star systems, the flat hollow disc picture is *much* too simple. And recall that these systems are not occasional oddities, but very common – possibly involving more than half of all stars. Planets that orbit one of the two stars in a binary can have stable orbits, providing that they are much closer to their host star than to the other one. For planets with circumbinary orbits, the condition for orbital stability is in a sense the opposite. Now the distance between the two stars needs to be smaller than the distance between the planet and either of them. However, we do not yet know whether the probability of planets forming in the first place is the same in binary systems as in those with just a single sun. It might yet turn out that the probability is significantly lower.

In more complex systems, everything again depends on relative distances. The planet Proxima b has no problem in stably orbiting Proxima Centauri, because the other two stars, Alpha Centauri A and B, are so far away. And computer simulations suggest that stable orbits around A and B individually should be possible too. The middle of the habitable zone of star A is thought to be about 1.25 AU out, while that of the smaller star B is at about 0.75 AU. These distances are much smaller than the average A-to-B distance of more than 10 AU. So a habitable planet orbiting either A or B is a possibility. In contrast, there are some binary and multiple systems in which the distances apart of the constituent stars are such that orbital stability zones and habitable zones don't overlap at all.

Returning to the simpler situation of a single central star, there's another quasi-circular entity that we need to consider in addition to the habitable zone itself: a planet's orbit. We saw in the previous chapter that there is variation in how far from being circular these are, with Mercury's orbit being much less close to circular than that of the Earth, and comets' orbits being extremely elongated, or eccentric. It's much harder, of course, to measure the eccentricity of a distant exoplanet than it is for our local planets, but we do now have substantial data on exoplanet eccentricities, as can be seen from perusing NASA's *Exoplanet Archive*. There may well be some planets whose combination of average distance from their star and eccentricity of their orbit is such that they are in the habitable zone for only part of their orbital journey (Figure 10). The evolution of life on any such

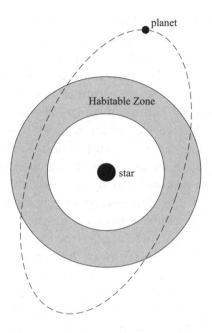

**Figure 10** A planet with a highly eccentric orbit that is sometimes within the habitable zone of its host star, sometimes not. Habitable zones themselves are always expected to be approximately circular in the case of single stars like our Sun. Planetary orbits, in contrast, are always (since Kepler) expected to be eccentric to some degree – that is, they are elliptical rather than circular. The degree of eccentricity varies markedly from one planet to another.

planet is less probable than on one whose whole orbit is entirely within the habitable zone, but it's not impossible. Much would depend on the proportion of each orbit during which the planet is outside the zone, and also on the nature of the planet itself – for example a global ocean (as on Earth) might be able to retain heat long enough, and release it slowly enough, to enable a planet spending a short period beyond the zone to remain habitable throughout its year.

So far in this chapter, we've seen that the idea of a habitable zone is very important, but not as simple as it first appears. And in the rest of the chapter we'll see further complexities. The one that we turn to next concerns the habitable zone's duration in time.

## Changes over Time

Everything changes, given enough time; and there is no reason for habitable zones to be exceptions to this general rule. So at this point we need to ask why they change and how much they change. Also, from the perspective of a particular planet, we need to ask whether the changing location of the habitable zone around its star has the result that the planet moves into or out of the zone. And from the perspective of living organisms, we need to ask whether life can be sustained if an initially inhabited planet suddenly finds itself no longer as habitable as it was in the past. Although this situation might arise due to a planet's orbit changing its location over time, most orbital migration probably occurs during the early stages in the history of a planetary system, before the evolution of life has started. So let's ignore this possibility and focus instead on changes in the star.

Our Sun is thought to be significantly hotter and more luminous now than it was at the time of the origin of life, some four billion years ago. And, four billion years into the future, it will be significantly hotter again – probably hot enough to boil away all surface water. If that happens, then by definition the Earth will no longer be in the habitable zone. But we are not talking about the red giant phase here – that comes later. Rather, these changes are within the main-sequence phase of the Sun – that is, the phase it's in right now and in which it will spend about 90% of its total lifespan. The phenomenon of gradually getting hotter and more luminous throughout this phase is not unique to the Sun – rather, it is a general feature of stars. But why does it happen?

During its main-sequence lifetime, a star is using up hydrogen as it converts it into helium through nuclear fusion. As this fusion proceeds, the ratio of hydrogen to helium keeps changing, and as a result the core of the star gets denser. The denser it gets, the hotter it gets too; and the higher the temperature is, the faster nuclear fusion works, hence increasing the star's luminosity. This process is a gradual and continuous one, in contrast to events that happen before and after the main sequence, such as the switching on of fusion in the first place or the explosion of a supernova. On a timescale of millions of years the gradual increase in luminosity really doesn't matter. The luminosity of the Sun now is negligibly different from what it was when the proto-human and proto-chimp lineages separated. But because the effects of increasing luminosity on a timescale of billions of years *are* significant, it will be the reason that the Earth eventually leaves the habitable zone – unless some unanticipated event intervenes in the meantime.

The Sun's main-sequence lifespan has been estimated to be about 10 billion years – so it is now just under halfway through that period. The length of time that it took from the origin of the solar system to the origin of life was about 0.5 billion years. The period from the origin of the system to the origin of animals was between 3 and 4 billion. For the origin of humans it was 4.5 billion. This last figure might seem odd as it's the same as the estimated age of the Earth. But in fact that's not odd at all – it's just a consequence of rounding off the numbers. The origin of *Homo sapiens* took place between 1 million and 0.1 million years ago. In billions, that's between 0.001 and 0.0001. When we're using a scale of billions, and rounding to the nearest 0.5, or even the nearest 0.1, the origin of humans was 0.0 billion years ago.

These ballpark figures for the time that it took to evolve microbes, animals, and humans on Earth can be used to estimate the maximum mass of star that a planet could orbit and manage to evolve life of these three types. To evolve humans, or indeed animals, a planet would need to orbit a smallish star belonging to one of the classes F, G, K, or M. That is, something from a little more massive than the Sun down to a red dwarf. Since there is a range of masses, luminosities, temperatures, and lifespans *within* each class, some of the F-class stars will have too short a lifespan to evolve humans or animals. This means that there is a kind of symmetry of stars able and unable to host the evolution of animals or humans, as follows: classes O, B, and A are all unable, classes G, K, and M are all able, and the central class F is split. For microbes, only half a billion years is required, so the dividing line falls not within class F but within class A.

There is of course a caveat here. We are estimating the time 'required' for the origin of three types of life-form from the actual time that it took to evolve them here on Earth. This is a pragmatic choice – we have no other information to go on. But it carries with it the risk of being wrong. Perhaps evolution can proceed much faster elsewhere than on Earth. There is no reason to suspect this on the basis of what might be called deterministic processes. For example, there is no reason why evolution should progress faster on a planet orbiting an ultra-hot, ultra-luminous O-type star. To be in the habitable zone of such a star, the planet has to be much further out in its system than Earth is in ours, as we've already seen, so in a sense their locations are comparable. However, certain important evolutionary steps may occur sooner on some habitable planets than on others, simply due to historical accidents. For example, the two-billion-year wait for complex eukaryotic cells on Earth might be replaced by a two-million-year wait on another Earth-like planet, with the result that the origin of animals might also occur much sooner there than it did here. We'll look further at this and related possibilities in Chapter 13.

The appropriate extension of the habitable zone concept to deal with periods of time rather than moments in time is the continuously habitable zone, or CHZ; the term was first used by American astrophysicist Michael Hart in 1979. The CHZ is the zone within which, over a defined timespan, liquid water can exist on a planet's surface without interruption. Clearly, the longer the timespan, the narrower the CHZ will get; and over the entire lifespan of a Sun-like star there may be no CHZ – or, to put it another way, the width of the CHZ has reduced to zero. Although the main factor governing change in the habitable zone (and hence in the CHZ) over long periods of time is the increasing luminosity of a star during the course of its main-sequence lifespan, other factors may be important too.

The CHZ concept is clearly of particular interest to us in terms of the continued habitability of Earth. As the Sun continues to heat up, and the Earth does likewise, many issues will arise that are likely to adversely affect life, and they won't all happen together. The most obviously terminal change for all forms of terrestrial life will be the loss of Earth's water into space. Ken Caldeira and James Kasting have estimated that this will occur between two and three billion years from now. That gives us plenty of time – well, potential time anyhow – before the Earth becomes uninhabitable. If we make it that far, we might have the technology to colonize the closest suitable exoplanets, or to protect Earth with some form of sunshield, or both. On the other hand our technology, in the form of thousands of nuclear weapons that may one day be used, or millions of engines burning fossil fuels and driving a runaway greenhouse effect, may kill us off long before then. This uncertainty over our future nicely illustrates the combined positive-and-negative potential of technological advance.

## Life on Mars?

The increasing temperature and luminosity of a star over its lifespan means that the habitable zone moves progressively outward with time. In the context of our solar system, this means that in the distant future Earth will no longer be in the zone, as noted above. But by that time Mars might be in the zone, and might be able to evolve life in and around sizeable seas produced by the melting of its polar ice-caps – seas much larger than the body of liquid water discovered recently under the ice of Mars's southern polar region. There may even one day be a scenario where most of Mars's low-lying northern hemisphere is covered by a vast ocean, leaving the southern highlands as the only dry land.

Given this time-trend of outward movement of the habitable zone, we can also look back into a planet's past. Mars should have been even further outside of the habitable zone in the distant past than it is now. And yet, strangely, past life on Mars is often discussed as a serious possibility. The reason for this is twofold. First, the lower amount of heat received from the Sun may have been more than compensated for by Mars having a thicker atmosphere in the past. A thick atmosphere can cause a runaway greenhouse effect that involves the surface being at a much higher temperature than it would be otherwise. This is why present-day Venus has a higher average surface temperature than Mercury, despite being further out from the Sun. Second, there is evidence from various space probes sent to Mars – flybys, orbiters, and landers – that there was liquid water on the surface in the past, even though there is none now; and some of the probes even seemed to find evidence of life.

The first spacecraft to attempt landings on Mars were the Soviet *Mars 2* and *3* missions, both launched in 1971. The former crashed on Mars; the second achieved a soft landing, but contact with it was lost after just a few seconds. So these missions failed to provide any information on possible Martian life. However, just five years later, two US spacecraft – *Viking 1* and *2* – landed successfully and began to conduct experiments and send back data. Of particular interest were the *Viking* biology experiments, designed to test for the presence of carbon-based microbial life.

Of the three *Viking* biology experiments, one was designed to test for carbon fixation, as carried out by autotrophic organisms on Earth, while the other two were designed to test for metabolic activity, including the production of gases such as carbon dioxide. Although there were positive results in both of these respects, control experiments that involved conducting similar tests on heat-sterilized samples of Martian substrate also gave positive results. While there is not a complete consensus on the appropriate conclusions to draw from these strange findings, most scientists now agree that no biological processes were involved. In other words, this was a case of false positives.

Since *Viking*, there have been several other US Martian landers and rovers: *Pathfinder–Sojourner*, *Spirit*, *Opportunity*, *Phoenix*, *Curiosity*, and, most recently, *InSight*. These have not attempted biological experiments. However, some of them have been involved in analysis of Martian rocks and 'soils'. One of the most important findings to come out of these missions was confirmation that there was once liquid water on the Martian surface.

There are some beautiful photographs, taken by some of the landers and orbiters, of what look like dried-up river systems on

Mars. However, evidence for running liquid is not necessarily evidence for running water. After all, the lakes on Titan are predominantly liquid methane. How do we know that the ancient river systems of Mars contained flowing $H_2O$, rather than any other liquid? If the effect of the outward migration of the habitable zone was overridden by a greenhouse effect of an ancient thick Martian atmosphere, liquid water would have been possible. And deposits of hematite – a form of hydrated iron sulphide that requires the presence of liquid water for its deposition – suggest that water was indeed present.

Assuming that the liquid on the Martian surface was water, there is a possibility – but only that – of there having been life on the red planet in the past. Connected with this possibility, analysis of a Martian meteorite discovered in Antarctica in 1984 initially suggested that it bore fossil evidence of ancient Martian microbes. The meteorite is designated ALH 84-001, with the ALH standing for Allan Hills, the discovery site. This object is about four billion years old, and thus was formed around the time that life began on Earth. Fossilized remains of putative Martian microbes could have been incorporated into the rock as it formed, before it was subsequently blasted off the surface of Mars by an impactor arriving from elsewhere. Fossil microbes were claimed to exist in ALH 84-001 on the basis of photomicroscopy revealing small entities superficially resembling bacterial rods and filaments. However, the size scale doesn't match (the 'microbes' are too small). Also, given the scarcity of microbial fossils on Earth from 3–4 bya, the chances of finding some in a single piece of Martian rock about the size of a potato are remote at best. The most sensible conclusion, as in the case of the *Viking* results, is that this is a false positive, and we are seeing the effects of non-biological chemical processes.

Another observation that has sometimes been claimed to show the existence of biological activity on Mars is the presence of methane in its atmosphere. Since this compound is broken down rapidly by photochemical reactions, its continuing presence at detectable levels must mean that it's being produced somehow. However, while biological processes can produce methane, so too can various non-biological ones. Importantly, the mineral olivine can interact with water and carbon dioxide to produce methane; and olivine is common on Mars. Thus the most parsimonious hypothesis to explain the ongoing production of Martian methane – and also seasonal variation in Martian methane levels – is one that involves only chemistry, not biology.

To conclude, all lines of evidence for putative Martian life, both past and present, are open to interpretation in purely chemical ways. This doesn't mean that there is not, and never has been, life on Mars. But it does mean that anyone making such a claim in the future needs to be more sceptical of their own findings than the claimants of the past, and less willing to jump into press conferences or publications. Unless and until more convincing evidence is found, I will continue to be a non-believer in Martian life.

## Life Outside the Zone?

The above thoughts on the Martian situation lead to a refinement of the idea of a habitable zone. We might imagine two types of zone: in one, only the star's characteristics are taken into account to estimate how far away from it liquid water could exist on a planet's surface; in the other, aspects of the planet itself are also taken into account. These would include its atmosphere and its mass – upon which atmospheric retention depends. However, despite this complication, there are some parts of a planetary system in which liquid water could not exist on a planet's surface regardless of the thickness and composition of its atmosphere. These barren zones are located, as might be expected, very close to the star and very far from it.

In our own system, there is no doubt that liquid water could not exist on the surfaces of any of the bodies that orbit further out than Mars. Although the four planets beyond Mars don't have surfaces as such, their moons do. And one of these – Saturn's moon Titan – also has a thick atmosphere, as we noted in Chapter 3. However, this doesn't mean that it's heated by a runaway greenhouse effect – it's much too far out from the Sun for that to be possible. Its average surface temperature is less than 100 kelvins (below minus 170 degrees Celsius, or minus 270 degrees Fahrenheit). But there's a possibility that both water and life might exist below the surface of some of the outer solar system's icy moons, in particular Jupiter's Europa and Saturn's Enceladus.

We don't have any lander-based information on either of these moons. In contrast to the multiple landers and rovers sent to Mars, only a single landing has been made beyond 4 AU from the Sun – the landing of the *Huygens* probe on Saturn's largest moon Titan in 2005. This was part of the highly successful *Cassini–Huygens* mission, a joint venture between NASA, the European Space Agency (ESA), and the Italian Space Agency (ASI). Although *Huygens* was not equipped to look for life, it was able to make various observations and measurements of the atmosphere of Titan, and of the surface layer at the spot where it landed. Photographs taken by *Huygens* show a truly barren landscape. Although we know of lakes of methane else-where on Titan's surface, we don't yet know of any chemical processes that would enable life to exist in this liquid. Strangely, although Europa and Enceladus have neither atmospheres nor surface lakes, they are currently thought to be better places to look for possible life. Indeed, they are thought to be the most likely places for extraterrestrial microbes to exist in the solar system. The evidence for sub-surface lakes or oceans in these two moons is as follows.

The surface of Europa looks like a huge ice shell that has been fractured in many ways. Spectroscopic analysis of the surface reveals it to be water-ice with a variable amount of salt. The most broken-up parts of the surface present an appearance that is referred to as Conamara chaos – named after the region of County Galway in Ireland that is called Conamara (or Conne-mara in the British spelling). This region of Ireland has a sort of patchwork landscape of many small lakes dotted across a moor-land environment, with the opposite sort of patchwork – small islands dotted across a shallow sea – often found just off the coast. On Europa, the patchwork takes the form of ice-rafts that have been broken up, moved around, and re-frozen. It is thought that underneath the ice is either a series of lakes or a global ocean. The latter is rendered more likely from modelling work published by Margaret Kivelson and colleagues in the year 2000. Their models were based on results obtained from the magnetometer of the *Galileo* spacecraft, whose mission began in 1995. The models were consistent with a moon-wide electrical conductor just below the ice, the most likely form of which is salty water. Plumes of water erupting through cracks in the ice were observed in 2018.

Enceladus has been known to have plume activity for many years. The *Cassini* probe flew through these plumes and analysed them. They were shown to be predominantly composed of water-ice, with small amounts of ammonia, carbon dioxide, and simple organic molecules like methane. A recent study by Frank Postberg and colleagues showed that there are more complex organic molecules too. As with Europa, the plumes from the surface of Enceladus are interpreted as being underlain by either lakes or a continuous sea. The plumes are due to the water of this sea venting through cracks in the overlying ice, with the water jets freezing into small ice crystals as they escape into space.

Whether or not there is microbial life in the sub-surface water-bodies of Europa or Enceladus will not be known until space probes land there and drill through the ice, which may be no easy matter as it is thought to be several kilometres thick in many places. So, for the moment, life on these moons remains just a tantalizing possibility. But before we leave this possibility hanging, there's one final question to consider: what is the source of their heat, given the remoteness of the Sun?

The answer is 'tidal heating'. We've already met the related process of tidal locking. In general, 'tidal' refers to the gravitational effects of one body on another body. The affected body is usually smaller, but not always – consider, for example, the tidal effects of the Moon on Earth's oceans. Tidal heating on moons such as Europa and Enceladus is caused by deformation of the interior of the moon resulting from variation in the effects of the parent planet's gravity. This deformation leads to friction, which in turn leads to heating. It is this tidal heating that keeps the sub-surface seas liquid; and it is this same heating that *may* provide quasi-equivalent features to Earth's hydrothermal vents, thus perhaps enabling a form of chemosynthesis to occur. Personally, I'm sceptical – but it's possible.

## A Galactic Habitable Zone?

In Chapter 3, we noted that, unless there are forms of life that are utterly beyond our imagining, there are definitely barren parts of the galaxy – for example stars themselves and interstellar space. Such places – and hence most of the galaxy – can be thought of as uninhabitable. However, some scientists have gone further than this and tried to define a galactic habitable zone, with the corollary that there is a galactic 'non-habitable zone'.

At its simplest, the idea is that too close in to the central bulge of the Milky Way or too far out in its halo would not be conducive to life for various reasons. Close in, there would be too frequent supernova explosions, because star density is high in the bulge. Far out, the main problem might be the low metallicity of the old stars that make up much of the galaxy's stellar halo. A low value of metallicity of a star is associated with a low value in its proto-planetary disc and hence low metallicity in the planets that form from it. Since all the elements that are important to life, hydrogen aside, are 'metals' under the astronomical definition, life as we know it could not evolve on a planet where the metallicity is very low.

The problem with the idea of a galactic habitable zone is that the underlying rationale is probabilistic rather than being based on a clear defining criterion, like the liquid surface water criterion for a circumstellar habitable zone that we've already discussed. It may well be less likely for life to evolve on a planet orbiting a star close to the galactic centre or far out at the fringe, but it's not impossible. Thus dividing the galaxy into putative habitable and non-habitable zones makes little sense.

It seems appropriate to end this chapter with another reference to what logicians and mathematicians call 'necessary and sufficient conditions'. If microbes are one day discovered in the sub-surface ocean of Enceladus, this will mean that being in the habitable zone around a star where water can exist as a liquid *on the surface* is not a necessary condition for life. Likewise, if a planet of a comparable age to Earth can be found that orbits in the habitable zone of its star but is devoid of life, this will confirm what we strongly suspect already – that being in the habitable zone is not sufficient, by itself, for life-forms to exist on a small rocky planet. Nevertheless, despite the fact that a habitable zone may prove to be unnecessary and/or insufficient, it's the best starting point in our search for life. We need to retain and refine it, not discard it.

# 11 OTHER HABITABILITY FACTORS

## Aspects of Earth

We are fairly certain that simply being in the habitable zone is not a sufficient condition for life, because a gas giant right in the middle of the habitable zone of a distant Sun-like star will not host life as we know it, and perhaps not life of any kind. The probability of life actually evolving on a planet that is in the habitable zone is influenced by various factors. We can look at such factors in relation to the Earth, and then try to extrapolate from here to elsewhere. However, we need to avoid coming up with a long list of Earth-based conditions that we then deem to be necessary everywhere else too. This is what Peter Ward and Donald Brownlee did in *Rare Earth*, and it led them to the conclusion – in my view an erroneous one – that complex life is virtually non-existent in the galaxy. With this caveat in mind, let's now run through a list of features of the Earth and the solar system that may or may not be crucial for the origin, evolution, and continued existence of life.

First, we're a smallish rocky planet – albeit the largest such planet in our solar system – with a surface that's about 70% water, 30% land. A mixture of marine, freshwater, and dry-land environments enhances biodiversity on Earth, as the species adapted to any one of these habitats are generally different from those adapted to the others. But does this mean that such a three-way mixture is necessary to evolve life in the first place? Probably not, given that life on Earth almost certainly originated in water rather than on dry land. Yet there are some concerns about whether a 'waterworld' with a planet-wide ocean would be able to host life. Personally, I don't see why the lack of dry land should be a problem, at least for life to get going, though clearly a tree of life that evolved on such a planet would be very different from our own.

Second, our planet is embedded in a particular type of system. Our closest planetary neighbours are Venus and Mars, planets not too different in mass from the Earth. Beyond Mars, the next planet out is the ultra-massive gas giant Jupiter. It's sometimes said that Jupiter is essential to our continued survival because its huge gravitational attraction mops up many stray asteroids and comets that might otherwise have come in our direction. However, Jupiter's effect on such bodies is more complex than a generalized 'mopping up'. Under certain circumstances, Jupiter can sling an asteroid into an Earth-crossing orbit, as discussed by James Kasting in Chapter 9 of *How to Find a Habitable Planet*. In other systems, the balance between helpful and hindering effects (to life) of large planets may be different.

Then there's the Moon. It's unusual in terms of how big it is relative to the Earth. Although Jupiter's moon Ganymede is the local 'king of moons' both in mass and in volume, our own Moon has by far the biggest mass relative to its host planet, out of any of the eight planets in our system. This is important in at least two ways. First, it gives the oceans of the Earth pronounced tides. Without it, there would still be tides – caused largely by the Sun – but they would be much less dramatic. Second, the Moon's orbit around the Earth may act to stabilize the Earth's tilt, which currently has a value of about 23 degrees. However, neither pronounced tides nor a particular tilt – giving a particular pattern of seasonality – would seem to be essential to evolve life.

Deep underground, Earth is a very geologically active planet. We took a brief look at plate tectonics in Chapter 4. Associated with the movements of the plates are volcanic activity and earthquakes. Both of these can be seriously detrimental to life, as we're well aware. However, plate movements have beneficial effects too. The deep-ocean vent communities that we looked at in Chapters 5 and 6 would not exist without the geothermal energy that is associated with Earth being geologically active in general and having plate movements in particular. And although volcanoes pose a threat to life, they also provide a boost to the evolutionary process. Archipelagos of volcanic islands such as Galapagos and Hawaii are particularly fertile areas for the origins of new species – with Darwin's finches providing the best-known, but by no means only, example.

The above effects of plate tectonics are readily apparent when we think in terms of physical processes. However, other effects emerge only when we think also in terms of chemistry – and in particular the carbonate–silicate cycle through which carbon is transferred on a long-term basis between Earth's atmosphere, terrestrial environments, oceans, a variety of organisms, and the uppermost part of the Earth's mantle. We can characterize this cycle as follows. Carbon and water react in the atmosphere to produce a weak solution of carbonic acid ($H_2CO_3$). When this falls as rain on carbonate rocks such as chalk (think of the White Cliffs of Dover) and other forms of limestone, it causes weathering, and more carbonic acid is produced. This runs off the land and ends up in the sea via streams and rivers (or more directly in the case of marine cliffs). Many marine organisms, both large (molluscs) and small (the amoeba-like foraminiferans), incorporate this carbon into the calcium carbonate ($CaCO_3$) of their shells. When they die, their shells get crushed, and the resultant debris forms calcium carbonate sediments that become incorporated into the sea-floor. But since the sea-floor is part of a plate, it doesn't remain in one place; rather, it eventually gets subducted down into the mantle. At the high temperatures prevailing there, the calcium carbonate is involved in a series of reactions that liberate $CO_2$, which is sent back into the atmosphere via volcanoes, thus completing the cycle.

As well as involving *some* organisms – notably those with shells of calcium carbonate – the carbonate–silicate cycle is important for *all* organisms in that it acts as a long-term stabilizer (or negative feedback loop) on the climate. If conditions become very cold, there is less evaporation of water into the atmosphere, hence less conversion of $CO_2$ into carbonic acid.

So the removal rate of $CO_2$ from the atmosphere is lower. But the volcanic input rate remains the same, due to the long time-lag involved in the cycle. So the level of atmospheric $CO_2$ goes up, and so does the greenhouse effect; thus the temperature rises. Alternatively, if conditions become very warm, evaporation of water from the oceans goes up, and thus the removal of $CO_2$ from the atmosphere increases, while its input again remains the same. Thus the level of atmospheric $CO_2$ goes down, as does the greenhouse effect, and likewise the temperature. On a planet without plate tectonics, the carbonate–silicate cycle and its climate-stabilization effect wouldn't work as they do on Earth.

Geological mobility much further down than the plates is what causes Earth's strong magnetic field. As we saw in Chapter 4, it is movement of the liquid outer core, composed of iron and nickel, that's responsible for generating the magnetic field that we usually take for granted. Yet we shouldn't do so. Venus has no magnetic field, while Mercury and Mars have only very weak ones. Jupiter has a much stronger magnetic field than Earth; it's thought to be generated by movements within a liquid part of its core – but this time the liquid is 'metallic hydrogen' rather than a broth of molten iron and nickel.

Overhead, we know there to be a thick atmosphere that is essential for present-day life. And it's important in more than one respect. For aerobic life – which includes most current life – its chemical composition, and especially its high level of oxygen, is crucial. But what we might call its physical, as opposed to chemical, composition is important too. The fact that it is quite thick, not unbreathably thin as on Mars, nor thick enough to blot out the Sun as on Venus, is crucial. Also important is its possession of a stratosphere containing an ozone layer which helps to shield life from damaging radiation, as we saw in Chapter 4.

With regard to long-term continuation of life, we looked at the all-important spectral class of the Sun in Chapter 9. However, as well as being a G-class star with a long lifespan, there's another feature of the Sun that's important to life on Earth – it's *not* a flare star or a superflare star. We're accustomed to the idea that the Sun has frequent surface phenomena that affect our atmosphere, causing effects that are collectively known as space weather, including the disturbances that occur in our magnetosphere and are called geomagnetic storms. The two main types of solar surface phenomena are solar flares (energy) and coronal mass ejections (CMEs – mixtures of energy, matter, and associated magnetic field). These can occur independently or together. They are quite frequent events; they occur on a daily or weekly basis. Their frequency varies with an 11-year period; the solar cycle has a gap of approximately 11 years between one peak of activity (sunspots, flares, CMEs, etc.) and the next. Interestingly, if we include the Sun's magnetic field in the picture, the periodicity of the solar cycle changes from 11 to 22 years, because the Sun's magnetic field flips over every 11 years. Anyhow, the flares and CMEs that occur on the Sun are generally not a threat to life on Earth, even during the most active phases of the solar activity cycle.

However, some stars have flares that are much more energetic than those emanating from the Sun. We now recognize flare stars and superflare stars, both of which have potentially life-endangering stellar outbursts – with up to thousands of times the energy of a typical solar flare. Proxima Centauri is a flare star. We saw earlier that it's a red dwarf, and in general the stars that are categorized as flare stars are within the red dwarf class. An example of a superflare star can be found in the constellation of Aquila the eagle – Omicron Aquilae. This is rather similar to our Sun, and indeed all superflare stars characterized to date belong to the spectral classes F and G.

We now return to the point with which this section started: being in the habitable zone is not in itself sufficient for actual habitation. A planet that is tidally locked to a flare star will have one sort of problem. And a planet orbiting in the habitable zone of a Sun-like star with huge adjacent asteroid belts immediately inward and outward of it will have another. But neither of these problems need be fatal. Also, many of the features of the Earth itself, including its disproportionately large Moon and its particular strength of magnetic field, may be unnecessary for life.

## A Place for Life to Start

Under the Terraspermia hypothesis adopted here – as opposed to the alternative Panspermia hypothesis – life originated somewhere on Earth. We don't know exactly when, but it was probably between 4.3 and 3.7 billion years ago. We don't know where either. In terms of a precise point on the globe, we probably never will. But in terms of the type of habitat, we can at least narrow this down from all known past and present habitats, of which there are many, to a few – maybe even just to two of them. The two most favoured possibilities are that life originated either in warm shallow water at the fringes of an ocean or in hot deep water in the vicinity of hydrothermal vents.

As we saw in the first chapter, the origin of systems that could be called alive from collections of organic molecules such as sugars and amino acids – which are clearly not alive – involves the origins of macromolecules, metabolism, membranes, and reproduction. And that's the most likely temporal order of the events, though others are possible (e.g. reproduction before membranes), as is a messy process in which there are no clear stages. The total time taken for this key proto-biological phase of the origin of life is completely unknown, and might be measured in centuries, millennia, or millions of years – perhaps the last of these is the most likely.

Recall that the foregoing phase (i.e. pre-macromolecules) involves the chemical production of sugars and amino acids from carbon, hydrogen, oxygen, and nitrogen. This phase occurred (and continues to occur) all over the universe, as the presence of organic molecules in meteorites and interstellar clouds of gas and dust makes clear. And it took a very long time. The first hydrogen *nuclei* – alias protons – are thought to have formed shortly after the Big Bang, when the universe was less than a billionth of a second old. But the first hydrogen *atoms* – in which the protons were joined by electrons – weren't formed until about 400,000 years later. Carbon wasn't formed until the dying stages of the very first stars, perhaps a few tens of millions of years later again. And the simplest organic molecules, such as methane, followed by more complex ones such as sugars, were formed an unknown period of time after that.

So, we know that the pre-biological generation of molecules like sugars and amino acids took millions of years, and was not restricted to Earth. The proto-biological generation of the first cells from a starting point of sugars and amino acids may have taken a similar amount of time, but was probably restricted to the surfaces of the Earth and other similar rocky bodies orbiting within the habitable zones of high-metallicity stars. In the context of our galaxy, or the universe as a whole, the Terraspermia hypothesis (the local version) is replaced by the Autospermia hypothesis – that every planet with life experienced an origin of life *in situ*, rather than being seeded by space-borne spores from another world. Maybe in some planetary systems there are two Earth-like planets so close together that the likelihood of an interplanetary propagule is non-negligible (like the TRAPPIST-1 system which we'll discuss in Chapter 14); but even in such cases my guess would be that independent origins of life on each planet are more likely.

Whether I'm right about this depends on the answer to a broader question: is the origin of life on the surface of a habitable planet something that follows naturally from physicochemical processes that take place there, and hence is almost inevitable, or is it dependent on some fluke event or one-off historical accident, and hence is vanishingly unlikely even in this most promising of environments? This question is posed by Addy Pross in his chapter on the origin of life. I agree with him that the former is more likely than the latter, and indeed this view was embodied in the Mosaic hypothesis of Chapter 3. But some other scientists take the opposite view.

Another useful distinction made by Pross is that between historical and ahistorical approaches to the origin of life on Earth. Taking the historical approach, we try to answer questions involving particular details, such as whether life arose in 'warm shallow' or 'hot deep' marine environments. Taking the ahistorical approach, we ignore such details, and grapple with the ways in which a mixture of organic molecules and water could become a system of proto-cells in general. From the perspective of assessing the potential habitability of another planet, it is the ahistorical approach that is the more important of the two. The details of one origin of life will undoubtedly differ from those of another; but the broad sequence of events may well be similar.

Just as the origin of life probably differed in detail from one inhabited planet to another, so too did the resultant earliest life-forms. Related to this, there is a lack of clarity about whether the first Earthly microbes were extremophiles or mesophiles, and about whether they were more similar to present-day bacteria or archaea. The answer to the first of these questions does not help with regard to answering the second, since we now know that both of these domains of life include members that are extremophiles and members that are not. We shouldn't declare a planet non-habitable because it lacks deep-sea vents, because, even if one of these was the location of the origin of life on Earth, this does not imply that a similar type of environment was necessary for life to originate elsewhere.

## The Continuation of Life

For a planet to be habitable in a meaningful sense, it needs to provide not only the conditions under which life can originate but also those under which it can survive and prosper. In other words, although such a planet will, like all others, have a varying rather than invariant set of surface conditions, the pattern of variation must be constrained in such a way that it is survivable through time. In the present section, we look at the short-term aspect of this problem – in particular, variation in environmental conditions that takes place on timescales of up to a year. We then look at its long-term counterpart (in the following two sections); and we end this chapter by looking at variation in space rather than time, in the context of a tidally locked planet.

On Earth, organisms experience various types and timescales of short-term environmental variation. Today at 3.15 pm a small invertebrate walking across my lawn would have been almost free from the danger of attack from the sky, but at 3.16 pm it was subject to a dense storm of incoming missiles that were larger than it and made of solid ice – unusually large hailstones. This sort of change is very abrupt – it's measured in seconds or minutes. Changes that are part of a regular cyclic pattern do not usually occur on such short timescales, but some of them are not far behind in this respect. The difference between being submerged in water and being exposed to the drying power of the air is of huge importance to an intertidal organism, and happens on a timescale of hours. On a timescale approximately double that of the tidal cycle, the difference between night and day is crucial for a nocturnal predator such as an owl.

Nothing in a completely natural habitat happens with a periodicity of a week, as that's a human invention. A month, however, is based loosely on the period of the Moon's orbit. This gives the periodicity of spring and neap tides, each of which occurs twice in a lunar month. And when we get to a periodicity of a year, there are seasonal patterns galore, including the pronounced hot–cold cycles of all temperate regions, and the wet–dry cycles of many tropical ones. Even the arctic tundra has seasonality. To find a near lack of seasonality, we need to look deep down below the surface. Seasons are much less apparent at a great depth in the soil than they are in the leaf-litter layer. And those amazing hydrothermal-vent systems are only seasonal in terms of the organic material falling down on them from above.

Nocturnal owls and the invertebrates crawling across my lawn are Earthly creatures; their extraterrestrial counterparts will be subject to different environmental regimes. On another planet, a day could last for many Earth months; a year could last for several Earth decades. Alternatively, a year on a particular planet could last for less than a day there – as in the case of Venus. Seasons could be non-existent – if there is no planetary tilt – or much more pronounced than our own. There may be no large oceans. If there are oceans, they may have no tides at all, or a complex pattern of tides caused by multiple moons. To what extent would these possibilities inhibit or adversely affect life?

It's popular to emphasize the fragility of ecosystems on Earth. And it's true that some of them are fragile, especially given human onslaughts. The bleaching of coral reefs is a good example. Yet life in general is very resilient on our planet. The bare rock of a quarry floor can in time become mature forest. I drove down a road in County Mayo recently where the central tarmac had disintegrated and a linear grassy mini-forest had grown up to the point of impeding the progress of my car. The spoil heaps around sites of the mining of heavy metals such as lead and zinc in the nineteenth century at various locations in England are now covered with plants that have evolved to cope with normally toxic levels of these substances. The forest-covered crumbling cities of the future beloved by sci-fi movies may well be real one day – though if there's a high level of ionizing radiation that will be more of a problem for the plants than growing on dilapidated buildings.

In terms of life on other planets, we should be hesitant in leaping to a hasty conclusion that lack of Earth-like patterns of short-term variation would be a problem. Tides and seasons may be unnecessary for life. And if they exist in extreme forms, that may be fine also, at least up to a point. The two biggest threats to the survival of life on planets that have different patterns of short-term variation than the Earth are probably the lack of a day–night cycle and the lack of a year in which the planet remains continuously in the habitable zone. In astronomical terms, these two threats are associated with tidal locking and highly eccentric orbits respectively. We dealt with eccentric orbits in Chapter 10; we've also met tidal locking (in Chapter 9), and we'll deal with it further in the final section of this chapter.

Our conclusion so far from a consideration of the possible impacts on life of variation in environmental conditions over short periods of time – ranging from seconds to a year – is that life on Earth is remarkable resilient, and there is no reason to believe that life elsewhere would be less so: this is the Resilience of Life hypothesis. Terrestrial life has evolved to deal with all sorts of short-term environmental variation, both irregular and regular in its occurrence, with the latter including tidal, diurnal, and seasonal patterns. Extraterrestrial life has probably evolved on multiple planets across the galaxy and beyond to do the same – well, the same in a general sense, but with differences in detail. A bigger threat to life on Earth, and probably elsewhere, exists in the long term and involves events that are sporadic – events that occur on a timescale of millions of years.

## Long-Term Threats

The main type of *external* threat to an inhabited planet in the long term is the possible impact on its surface of a large body arriving from elsewhere in the planetary system concerned. Let's use the huge amount of information we have on this problem affecting the Earth; we can then use this as a pointer to what may happen in terms of impacts on inhabited planets in other systems.

On a clear night, an hour or so spent stargazing will probably reveal at least one 'shooting star', or meteor. These are tiny particles that were left behind by sublimating comets, which burn up as they are subjected to the friction of the Earth's atmosphere. Typically their size varies between that of a grain of sand and a pea. Because they are so small, nothing is left to hit the ground; friction-burning destroys the entire particle. Much less frequently, a larger body enters the Earth's atmosphere and, although part of its surface is lost to friction-burning, the central part survives its atmospheric journey and hits the ground. This happened near Chelyabinsk, in southern Russia, in February 2013. In this case, the body exploded at an altitude of about 30 kilometres (19 miles); it showered the area with multiple meteorites, the largest of which weighed 650 kilograms (about two-thirds of a ton).

In between tiny particles and huge rocks, meteorites that could be referred to as stones hit the Earth at a rate that's been estimated as at least a few thousand per year. Meteorites larger than the Chelyabinsk one luckily arrive rather rarely. In general, we can describe the situation as an inverse relationship between size and frequency: the bigger the size-class of the body, the less often impacts happen.

From the perspective of threats to life on Earth, what we're most interested in are the very largest impacting bodies. The largest meteorite ever found is the Hoba meteorite, which fell near the town of Grootfontein, in northern Namibia, about 80,000 years ago. Because it is so large, it is still located where it fell. With a weight of more than 50 tonnes (or tons), it makes the largest meteorite from the Chelyabinsk explosion look small. However, the Hoba meteorite itself is small when compared to the impactors that caused the largest known craters on Earth, at least one of which was involved in a mass extinction event.

The largest crater on our planet is the Vredefort crater, in the Free State province of South Africa. This crater is thought to have been originally some 300 kilometres (185 miles) in diameter, though it is now much eroded. It dates from an impact that occurred just over two billion years ago – at a time when life was represented solely by microbes. It may well have caused a mass extinction event, but the fossil record from that early time in the history of evolution is too poor for us to have conclusive evidence for such an effect.

The second largest crater – Chicxulub, in Mexico – most certainly was associated with a mass extinction. This crater has a diameter of about 150 kilometres (93 miles) and derives from an impact that occurred 66 million years ago – exactly the time when the dinosaurs became extinct. It is now generally agreed that the impactor involved threw up much debris and dust into the atmosphere, causing major climatic changes that were long-lived. These changes caused the extinction of about three-quarters of all animal and plant species, with the dinosaurs being the most famous casualties.

The Earth is unique in terms of the exact pattern of impactors it has received over time. But this uniqueness relates only to the details. Other planets, both in our solar system and elsewhere, will have experienced patterns that were similar to Earth's in *general* terms – for example, it's likely to be the case that small impactors are almost always more common than large ones. The relative commonness of different *types* of impactors, for example asteroids versus comets, may differ with position in the system.

But in general we can say that all inhabited planets face a threat to continued life from the impact of large rocky bodies on their surfaces.

What about *internal* long-term threats to life, that is, those that are intrinsic to the planet itself, rather than imposed upon us from space? Looking back at the Earth through geological time, a type of threat that occurs with a broadly similar frequency to the largest impacts is the occurrence of ice ages, of which there have been six so far in Earth's history. The earliest occurred about 3 billion years ago, while the latest was the recent Quaternary ice age, which ended a mere 10,000 years ago (or, using other criteria, is still continuing). In between were the Huronian glaciation (about 2.3 bya), the Cryogenian (about 720–630 mya, which includes the Sturtian and the Marinoan glaciations), the Andean-Saharan (450–420 mya), and the Karoo (360–260 mya). All involved extensive glacier coverage of land surfaces.

An interesting question is whether there has ever been a time when the entire Earth has been glaciated. There's a hypothesis to this effect – the Snowball Earth hypothesis – that refers in particular to two spans of geological time: the early Proterozoic (the Huronian glaciation, about 2.3 bya) and the late Proterozoic (Cryogenian, about 700 mya). However, it is really a spectrum of hypotheses ranging all the way from 'the most extensive glaciations of all but still short of total coverage' to a more extreme 'all of Earth's surface covered in thick ice'. It's hard to know exactly where the truth lies on this spectrum.

A typical ice age will affect life in the following way. Some low-mobility organisms restricted to small areas in temperate regions will become extinct. Other organisms with broader distributions and/or greater mobility will survive, albeit with their geographical ranges substantially altered. It is not thought that glaciations are major players in the causation of mass extinctions, though in some cases they may be contributory factors. However, if the more extreme version of the Snowball Earth hypothesis is correct, the Huronian and/or Cryogenian glaciations could have caused mass extinctions, particularly of taxonomic groups of land biota – though these would have been entirely microbial groups, as there were probably no plants (except algae) or animals at those early stages of Earth's geological history.

The causes of glaciations are unclear, and the factors involved may be different between one ice age and another. The Huronian glaciation was probably linked to the Great Oxygenation Event that we first met in Chapter 4. The nature of the connection is thought to have been as follows. As Cyanobacteria began to produce substantial quantities of oxygen, this was first taken up by various oxygen sinks, such as iron. After all the available iron had been oxidized (or rusted), oxygen began to accumulate in the atmosphere. As it did so, methane was oxidized to carbon dioxide, with the latter being a much less efficient greenhouse gas; this atmospheric change may have led to the Huronian glaciation. Although there was also a mass extinction event about that time, this was probably caused not by the ice age but by the new high level of oxygen in the atmosphere being toxic to many of the previously thriving anaerobic microbes.

## Mass Extinctions: Pattern and Severity

As well as looking at long-term threats to life, we can also examine the actual pattern of major declines in the diversity of life on Earth that we call mass extinctions. The severity of these gives us a clue about how close life on Earth has come to being wiped out completely.

Over the last 300 million years, there have been three mass extinction events, as follows. The biggest of all was the Great Dying, which happened at the end of the Permian period, 252 mya. More than 90% of species became extinct in this event. About 50 million years later, at the end of the Triassic period (200 mya), there was another mass extinction in which about 75% of species perished. And after another lull of more than 130 million years, the dinosaur-killing end-Cretaceous mass extinction event took place, again with about 75% of all species perishing.

Earlier extinction events are less clearly defined from the lulls that separate them than are the above three. At least two further events are recognized in the span of time from 300 back to around 500 mya. Going back before 541 mya, to before the start of the Cambrian period and thus before the start of an abundant fossil record, there were at least two more. One was *just* before the start of the Cambrian and involved the extinction of the enigmatic creatures known as the Ediacaran biota. The other was the much earlier mass extinction mentioned in the previous section that was associated with, and probably caused by, the Great Oxygenation Event.

The fossil record is not good enough for us to put a percentage extinction figure to either of these early events. And the estimated percentage extinction rates for the more recent mass extinctions have to be treated with some caution. However, if they are approximately correct, then the closest the Earth has ever come to having evolution stopped in its tracks was 252 mya in the Great Dying, when only about 10% of species survived.

Life's passage through this event sounds rather precarious, and in some ways so it was. But 10% of a very large number of species is still quite a large number. If there were approximately a million species of organisms 252 mya, there were still 100,000 of them 251 mya. Concentrating on the survivors rather than the casualties helps to emphasize just how great a potential remained for subsequent evolution. And indeed rapid evolutionary diversification of survivors is typically found to follow a mass extinction. As they say, nature abhors a vacuum.

This brings us back to the point made earlier about the resilience of life on our planet. What about life on others – should we expect it to be resilient too? I really don't see why not. The chances are that on other inhabited planets life has come through various times of crisis, just as it has on Earth. Of course, there may be some unlucky inhabited planets that were struck by such huge bodies that the effect was terminal – even to the point where the planet was shattered. But these are likely to be a small minority of cases. Our understanding of planetary systems is that, when they are mature, planet-shattering collisions are rare – although in the early stages of a system collisions that shatter *proto*-planets are considerably more common. Given that evolution probably doesn't usually get started until after the most dangerous phase of bombardment, the tree of life it generates should be capable of withstanding a certain amount of pruning; and the more branches it has, the safer it will be in this respect.

## Tidally Locked Planets

Having examined threats to life from temporal variation in the environmental conditions that prevail on a planet, we now look at threats from spatial variation. We will focus on one particular type of spatial variation, namely the extreme variation of temperature and other variables between one side of a planet and the other that may be caused by the phenomenon of tidal locking.

Recall that in a synchronous orbit caused by tidal locking, a moon will always present the same face to its planet, and a planet will always present the same face to its sun. A planet locked to its sun in the same way that the Moon is locked to the Earth is unlikely to host a biosphere like our own. However, here's an interesting hypothesis: it may be able to host a more restricted *bio-region*. One possibility of this kind would be a bio-ring, in other words a ring of inhabited territory around the border between the planet's lit and unlit halves.

From a biological perspective, bio-rings around tidally locked planets within habitable zones at first seem quite plausible because of the way evolution works. Just as natural selection is blind to time and cannot see into the future, so it is blind to space, and will not be prevented from occurring because the width of the ring in which it is happening is (say) just a few tens or hundreds of kilometres. However, from a climatic perspective, bio-rings are less plausible. For example, if the atmosphere transfers water to the dark side, where it freezes (impossible for the atmosphere-free Moon), a bio-ring would not be able to form. Then again, if the atmosphere is sufficiently thick, the dark side of a tidally locked planet might be kept above freezing, rendering the evolution of an extensive bio-region possible. Indeed, in this situation, the bio-region might take a shape that is somewhere between a ring and a spherical shell. However, photosynthetic life would still be precluded on the dark side.

How far might natural selection go in a restricted bio-region? On the one hand, this is as open-ended as the proverbial question 'how long is a piece of string?' – but on the other hand, it's not. There are many examples of spatially restricted places on Earth where evolution has happened over an extended period of time despite having little or no contact with other places; for example, remote islands. Although that scenario sometimes produces unusual creatures – like the 'hobbit', the human species of small stature that evolved on the island of Flores – having unusual results is not the same as being stopped in your tracks. And there is no reason why on another planet spatial restriction should prevent evolution from happening. The fundamental nature of natural selection would not be expected to vary between one planet and another – tidally locked or otherwise.

This is a very important point from the perspective of the commonness of life in the Milky Way and beyond, because tidal locking is common, not rare, as pointed out by the American planetary scientist Rory Barnes in 2017. The mechanics of the locking process are complex, but they are such that the zone of locking extends outward as the mass of the star increases; the habitable zone also moves out, but at a different rate (Figure 11). In our own solar system, Earth is well outside the tidal-locking zone, but Mercury is very close to it. Indeed, although Mercury is not completely tidally locked to the Sun, it exhibits a related phenomenon – spin–orbit resonance. Its year lasts 88 Earth days, while its day lasts for 59 Earth days – a 3:2 resonance factor. In this context, we should note that the effects of the two phenomena on habitability are likely to be very different from each other, despite their common origin in tidal forces. Complete locking poses a more severe threat to life than does a resonance.

The habitable zone is mostly within the locked zone for red dwarf systems, and partly within it for K-class systems. The

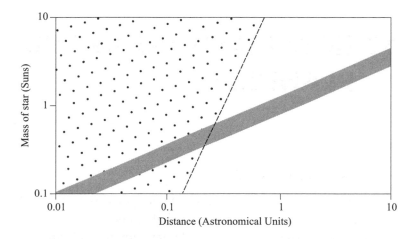

**Figure 11** As the mass of a star, and hence its luminosity, increases, the habitable zone around a star (shown in grey) moves outward. The maximum distance from the star within which tidal locking of planets takes place also moves outward (the tidal-locking zone is stippled). For a small red dwarf (mass 0.1), the habitable zone is entirely within the tidal-locking zone. For a G-class star like the Sun (mass 1) and all larger stars, the habitable zone is entirely outside the tidal-locking zone. For some K-class stars, the inner region of the habitable zone is expected to be within the tidal-locking zone. Note that both axes have log scales. Modified from Kasting (2010), with permission.

further we go towards massive O-class stars, the lower is the probability of a habitable-zone planet becoming tidally locked. This means that although the red dwarf end of the spectrum has the huge advantage of its stars, and hence planets, being long-lived, it has the disadvantage of maximizing the chances that its planets will be tidally locked and hence capable of supporting only a limited bio-region at best.

# 12 HOW MANY INHABITED PLANETS?

## Homage to Frank Drake

The American astronomer Frank Drake pioneered the SETI movement (search for extraterrestrial intelligence) in the 1960s. Among other things, he is famous for formulating the equation that bears his name. The Drake equation represents a way of estimating the number of alien civilizations in the Milky Way that are currently broadcasting radio waves into space in an attempt to make contact with civilizations on other planets. Drake introduced the equation in 1961 as a way of stimulating debate. He was aware of various complicating factors that suggest additional terms might need to be added to the equation to make it more realistic. Nevertheless, he gave us a good starting point in the quest to make astrobiology quantitative. We will now use variants of the Drake equation to quantify the commonness of life. This chapter uses the ballpark approach of Chapter 3 as a launch-pad and ends with a series of estimated numbers of planets with various types of life.

Our conclusion at the end of Chapter 3 was that of the trillion or so planets in the Milky Way, billions have microbial life and millions have animal life. The number of planets with intelligent life will be much lower than the number with animals. We didn't generate a ballpark estimate for this third figure in Chapter 3, but we'll do so in due course.

Frank Drake's approach was somewhat different from the one we used in Chapter 3. Instead of starting from the number of stars and planets in the Milky Way, he started from the rate of star formation, expressed as the number of new stars born per year. From this he progressed in stages to the rate at which broadcasting civilizations were formed. His estimate of the number of such civilizations that exist in the galaxy at present was simply their rate of formation multiplied by the average time for which they last (in terms of the number of years for which they continue to broadcast). For example, if one broadcasting civilization arises per million years, and they last on average for 10 million years, and if these figures don't change much as the galaxy ages, then there are about 10 of them broadcasting right now, including humans.

It's easier to use Drake's approach for life in general than for broadcasting life in particular, because in the latter case there are more unknowns. Naturally, from the SETI perspective, the emphasis is indeed on intelligent life-forms with advanced technologies. But from a broader astrobiological perspective the occurrence of all forms of life, from microbes to mammals, is of interest. Are mammals likely to be a repeatable feature of evolutionary processes on other planets? We'll address that question, and related ones about the repeatability of evolution, in the next chapter. For now we'll ignore what type of animal we're dealing with and will just think in terms of unspecified animals, and indeed organisms more generally, starting with microbes.

## How Many Planets with Microbial Life?

At this stage we need to examine the part of Frank Drake's equation that relates to simple microbial life as opposed to intelligent life. But before doing so, it's worth pausing to reflect on equations in general. While I was writing this chapter, or more accurately while I was taking a pause from writing it because the sentences weren't flowing as well as I wanted them to, I saw on the BBC website the sad news of the death of the famous cosmologist Stephen Hawking. This caused me to reflect on various Hawking-related matters, one of which concerned his best-known book, *A Brief History of Time*, which I read many years ago. The astronomer Simon Mitton, working as an editor for

Cambridge University Press during the period when Hawking was looking for a publisher, told him that for every equation he put into the book his readership would be halved – which is no doubt partly why there are no equations in it at all, except for the mandatory $E = mc^2$.

Perhaps this advice exaggerates people's tendency to jump to conclusions about the degree of difficulty of a book on finding an equation leaping out at them as they flick through a copy in a bookshop or online; but perhaps not. Either way, any author contemplating the use of equations should certainly bear in mind their power to reduce the readership of a book, even if the exact proportion of readers lost per equation is hard to quantify.

So, here's my attempt to sneak in an equation by stealth. What we want to estimate right now is the number of planets in the Milky Way that are currently inhabited by 'microbial life'. I use this phrase very broadly, to include all forms of unicellular life on Earth, regardless of which of the three domains they belong to, and all forms of life elsewhere in the galaxy that are broadly similar to these terrestrial organisms. We'll call the number of microbially inhabited planets $N$. To apply Drake's approach to coming up with an estimate of $N$, we need just five numbers, as follows:

1. the rate of star formation in the galaxy, in terms of number of new stars per year
2. the probability of orbiting planets arising in a stellar system
3. the average number of potentially habitable planets per system
4. the fraction of such planets on which microbial life originates
5. the length of time (in years) that life persists: time from origin to extinction

The version of the Drake equation that applies to the number of planets with microbial life is simply these five things multiplied together. To see it as an equation rather than as a list, it helps to use symbols for each of the five numbered points. Let's use $r$ for rate (no. 1), $p$ for probability (no. 2), $n$ for number (no. 3), $f$ for fraction (no. 4), and $t$ for length of time (no. 5). Our equation is then $N = rpnft$. That's shorter than some words, and easily lost in a paragraph – I hope. But the really difficult thing here is not learning to be comfortable with the equation (which is a very simple one as equations go), but rather coming up with reasonable values for its five constituent parameters. However, this process is at least a little easier now than it was a couple of decades ago, given the many recent advances in our knowledge of exoplanets.

So, we begin with the rate of star formation. The Milky Way is thought to be about 12 billion years old. Some of its individual stars have estimated ages that would put their birth somewhat earlier than that; but then again, some of them have estimated ages that exceed the age of the universe. This serves to emphasize that these various estimates have quite broad error bars associated with them. Perhaps the best solution is to think of the Milky Way as being $12 \pm 1$ billion years in age. If we take a middle figure from the range of size estimates for the galaxy (100 to 500 billion), we end up with 300 billion stars. If the rate of formation of these stars had been constant over the whole of the galaxy's history, it would work out as 25 stars per year. In reality, the rate has varied somewhat; and there are a few other complicating factors such as the absence of extinct stars from the calculation. Nevertheless, our answer at least gives an order-of-magnitude estimate. Let's say that the rate of star formation is probably somewhere between 1 and 50 per year.

The next parameter that we need a value for is the probability of one or more planets forming around a star. To estimate this probability, we need to look at the data acquired so far from exoplanet hunting. However, the problem with this information is that it takes the form of a series of minimum rather than 'most likely' numbers. As we saw in Chapter 3, the number of planets discovered in a system thus far ranges from one to eight, with so far just a single case of eight planets known apart from our own system – Kepler-90. At the time of writing, the approximate number of systems with each number of planets is as follows: 2500 systems with just one known planet, 450 with two planets, 125 with three, 50 with four, 15 with five, and 10 with six, seven, or eight planets. There are also many stars that have been investigated and appear *so far* to have zero planets.

However, these are all minimum figures for the systems concerned because they represent only the number of *confirmed* planets. In many cases, other planets are suspected, and in almost all cases other planets are likely. Our two most successful planet-detecting methods are the transit method and the radial velocity method. The first relies on detecting dips in light when a planet transits across the 'front' of its star from our Earth-based viewpoint. The second uses the forward and backward movement of a star – again from our viewpoint – that is caused by an orbiting planet; as the planet comes towards Earth, the star moves slightly backward, and vice versa. Both methods tend to overlook small planets. Also, both methods face problems in the case of planetary systems whose orientation is not approximately side-on to us, the transit method particularly so. Systems that are oblique in orientation will not exhibit transits, and thus most of these planetary systems go undetected. (This problem will be obviated by the next generation of direct-imaging space telescopes: see Chapter 15.)

What fraction of stars really have no orbiting planets at all? From what we know about the way in which stars form from collapsing clouds of gas and dust that lead to proto-stars and proto-planetary discs, the answer would seem to be 'a very low fraction'. It's hard to estimate exactly how low, but probably less than 10%, and possibly very close to zero. If that's true, then at least 90% of stars have some sort of planetary system, consisting of anything from one to eight or more planets. Let's use this as our best guestimate for the $p$-parameter of the equation.

Now we need to estimate the average number of potentially habitable planets per system. This number for our own solar system is at least one. If we consider our system to be typical, then 'one habitable planet' would work for other systems too. However, that's unlikely to be accurate, because we know of many systems in which there appear to be no habitable planets at all. Although we know of some systems where there's more than one planet in the habitable zone, such as TRAPPIST-1 (which will be discussed in Chapter 14), this is likely to be a smaller upward effect on the average than the downward effect of instances of zero habitable planets. Let's use a value of 0.05 for the $n$-parameter in the equation, meaning that one habitable planet occurs on average every 20 planetary systems. However, we should note that the true average value could well be up to an order of magnitude higher, at 0.5; and we should remember that underlying any such proposed average figure there is likely to be variation, for example between the different classes of stars, from O to M.

Next comes what is perhaps the hardest parameter of all to estimate in the slimmed-down version of the Drake equation that we're using for microbial, rather than intelligent, life. This is the fraction of habitable planets that actually become inhabited. My personal view is that it is close to 100% – for reasons articulated in the Mosaic hypothesis of Chapter 3 – but then again there are scientists who think it is close to 0%. So the best approach is to take a figure in the middle of the range and guestimate 50%.

The final parameter is a little easier to estimate, because we have some relevant data, if only from our own planet. The length of time that microbes have existed on Earth is about 4 billion years so far; and if they last until the Sun becomes a red giant – unlikely, given the probable loss of Earth's water long before then – they will endure for about 9 billion years overall. In other systems, microbes will last different lengths of time. One major constraint on their timespan is the spectral class of the sun in the system concerned. We've seen that at one extreme, systems orbiting O-class stars, evolution is unlikely even to have time to get started; and at the other extreme, M-class red dwarfs, evolution may potentially be able to continue for trillions of years. Since red dwarfs are the commonest type of star and O-class giants the rarest, microbial life will on average be longer-lived elsewhere than on Earth, providing it can survive the tidal locking that is common with M-class stars. Nevertheless, let's be cautious about this, and use our Earth-based figure of 4 billion years to date (ignoring possible future endurance), without applying an increase to take account of the skewed distribution of stars over spectral classes.

We can now do the requisite calculation to provide an estimate of the number of planets in the Milky Way that are currently inhabited by microbial life. Our estimated number is: $25 \times 0.9 \times 0.05 \times 0.5 \times 4$ billion. This works out at 2.25 billion planets. Our conclusion from this result is that microbial life is extremely common today in our galaxy, even though the fraction of planets it occurs on is less than 1%. The high number is exciting, but the low fraction urges caution regarding testability – it suggests that we would have to survey a very large number of local stars to find one planet with microbial life.

## Some Ifs and Buts

Now we should try to get some idea of how the estimate of 2.25 billion planets with microbial life is influenced by the errors associated with our estimation of the values of the various parameters of the microbial version of the Drake equation. We'll take them in the same order as we did in the previous section.

So, we start with the errors involved in estimating the rate of star formation. We used a value of 25 stars per year, but acknowledged that the actual rate could be anywhere from 1 to 50. If we were to use either extreme value rather than the compromise one, this would alter the estimate of the number of microbially inhabited planets to either 90 million or 4.5 billion. Also, we should note that our approach is based on the formation rate of *all* stars. However, there are stars that are too short-lived for the evolution of microbes, and stars that present complications for habitable zones, for example because they are in binary systems or are flare stars.

The proportion of all stars that have one or more planets was guestimated to be 0.9. But like all the other estimates, this figure could be either higher or lower. It seems clear at this stage in exoplanet science that the majority of main-sequence stars, or possibly even all of them, have planets. So the range of likely values is from 0.5 to 1.0. The former would almost halve our estimate of the number of microbial worlds to 1.39 billion, while the latter would increase it slightly to 2.5 billion.

The average number of habitable planets per system was guestimated at 0.05. It's hard to know the errors around this figure at present, though as we discover more exoplanetary systems it will get easier. Let's say that the true figure could be half or double our chosen one, in which case our estimate of the number of microbial worlds could be 1.125 billion or 4.5 billion.

Now we arrive back at the hardest parameter to estimate – the fraction of habitable planets that actually become inhabited by microbial life. We used 0.5 as an arbitrary figure in the calculations of the previous section. What happens if we use more extreme values? The problem here is 'how extreme is extreme?' Let's use 0.1 and 0.9. In this case, the estimated number of inhabited planets shifts to either 450 million or 4.05 billion.

Finally, we need to consider the errors surrounding the 4-billion-year estimate of the duration of microbial life on Earth. The errors on its duration so far are quite low – about 3.7 to 4.3 billion, as we saw in Chapter 4. But if we include the possibility of microbes surviving on Earth until the Sun begins to die, which doesn't seem unreasonable given the general resilience and rapid evolution of microbes, some of which might find patches of liquid water deep down in the ground even after all surface water has been lost into space, the maximum timespan for microbes on Earth is about 9 billion years. And microbes in many other planetary systems may already be much older than those on Earth. Consider a red dwarf system that began after the very first large stars had significantly increased the metallicity of its region of space – but not long after. Let's put a figure of 12 billion years ago on this. So, we can use 3.7 billion and 12 billion as our lower and upper bounds of the average lifespan of microbes on a habitable planet – to date. With these figures, the estimated number of microbially inhabited planets becomes 2.08 billion or 6.75 billion.

It's time to recap. In Chapter 3, using a ballpark approach not based on the Drake equation, we came up with a very rough guestimate of the number of microbial worlds in the galaxy of 'a few billion'. Now, using Drake's approach but simplifying his equation to deal with microbial life rather than intelligent life, we've arrived at an estimate of 2.25 billion planets. Having contemplated the likely errors around each of the estimated parameter values in the equation, our lowest estimated number of planets with microbial life is just under a tenth of a billion, and our highest number is just under 7 billion (the 'exact' figures being 0.09 billion and 6.75 billion respectively).

There's good reason to use 'exact' with inverted commas in the previous sentence. This usage is not exact in the sense that 1 metre is exactly equal to 3.28 feet. Actually, neither of these usages is exactly right, but the metre/foot conversion figure is a lot more exact than our microbial-worlds figure. So we should use the rounded figures, not the 'exact' ones. It probably makes sense also to round off our initial estimate of 2.25 billion to 2 billion. So the bottom line is this: the number of microbial worlds is estimated to be about 2 billion, and at any rate within the span of values from about 0.1 billion to 7 billion. But is this *really* the bottom line? To answer this question, let's make a brief digression from planets to human heights.

Imagine trying to estimate the average height for a reasonably homogeneous and well-defined group of humans – say the adult females of Iceland. The size of this group is more than 100,000, so we can't measure them all. But a random sample of just 100 women from the island will give a tolerably good estimate of the overall average. It's unlikely to be wrong by more than about a centimetre. And indeed the degree of uncertainty about our estimate can itself be calculated – it's usually given in the form of upper and lower confidence limits.

We need to admit that our guestimates of the number of microbial worlds in the Milky Way do not belong to the same realm of science as estimates of average human height or, to take another example, estimates of the average size of meteorites that derived from the explosion of that large chunk of asteroid over Chelyabinsk in 2013. Instead of having well-characterized errors, we have just a very rough idea of their magnitudes. What this means is that our guestimate for the possible range of the number of planets with microbial life in the Milky Way might be too narrow. Rather than thinking of it as being a predicted value of 2 billion within a range from 0.1 billion to 7 billion, we should perhaps think of it in only order-of-magnitude terms as being a predicted value of 1 billion, flanked by a very broad, but hard-to-specify, range of errors, extending at least from 0.1 billion to 10 billion.

It's taken a long time and a lot of guestimation to get here, but having arrived I think this is the real bottom line. The number of planets in the Milky Way that are hosts to microbial life right now is probably about a billion, working to an order of magnitude at best. As a fraction of all our galaxy's planets, this is about 0.1%. Truly, microbial life is both common and rare. But is the same true of 'complex' life? That's our next port of call.

## The Road to Animals

It's important not to see evolution as a ladder, staircase, or escalator of progress. In other words, it's important not to think in terms of a *scala naturae*, as noted in Chapter 8. Here on Earth, evolution can proceed from simple to complex or from complex to simple, with the latter trajectory being common in lineages that evolve from free-living to parasitic forms. And most evolutionary changes take the form of diversification – or 'becoming different' – *within* a level of complexity; that is, the descendant species are neither more nor less complex than their ancestors.

That said, it is unreasonable to expect evolution on an exoplanet to start with animals and eventually produce descendant microbes. At its beginning, any evolutionary process that starts from collections of organic molecules is constrained to produce only simple proto-unicells or something of that kind. Complex life-forms like animals must wait their turn. Whether they must wait as long as they did on Earth – over three billion years – is an open question. Perhaps in some cases they can evolve from microbes in millions of years rather than billions. Perhaps in some cases they never evolve at all. But where they do evolve, they can't come first.

What this means is that the number of planets with animal life is a subset of those planets with microbial life – unless there are some planets somewhere in which animal-type life survives while all types of microbes have become extinct. The probability of this happening seems vanishingly small, given the robustness of microbes.

We need to consider the relationship between animal life and 'complex life' in addition to the relation between animal life and microbial life. Recall that in the Rare Earth hypothesis, the type of life that Peter Ward and Donald Brownlee claim to be uncommon in our galaxy, and indeed in the universe as a whole, is 'complex life'. Although they equate this with 'animals and higher plants', as we saw earlier, they don't provide any discussion of higher plants at all. In fact, the only plants that they mention by name are red and green algae, which most botanists would consider to be lower plants rather than higher ones.

As multicellular eukaryotes, plants and animals have a lot in common. They typically have tissues and organs. They reach large body sizes. They're not unique in these things, because we can find some very large and quite complex organisms that don't fit into the animal or plant kingdoms – for example some fungi and brown algae. But some animal and plant groups do represent pinnacles of complexity in the context of life on Earth.

Here I'm going to concentrate on animals rather than plants, partly because, as a zoologist, I know more about them, and partly because they represent the realm in which (at least on Earth) intelligence can arise. However, much of what is said in this section probably applies to the plant kingdom too. In particular, multicellular plants must evolve from a microbial beginning, just as animals must, rather than having an evolutionary process that starts with – say – oak trees, and later produces microbial descendants. In terms of whether plants or animals came first on Earth, this depends on how we define both 'animal' and 'plant'. Red and green algae came before the first animals, while land plants came after them.

When did the first animals live on Earth? As with microbes, we can't specify an exact figure, but we can bracket the time of the first animal with a reasonable degree of certainty. Considering all types of evidence, including both fossil-based information and comparative molecular data on extant animals, it looks almost certain that the very first animal on Earth lived between 1.0 and 0.5 billion years ago. Now let's see if we can bracket it any more precisely than this (Figure 12). To narrow things down, we start in the fossil record of the Cambrian period – between about 0.54 and 0.49 bya, or, if you prefer, between about 540 and 490 mya.

One type of early animal that is easy to recognize is the trilobite – a type of marine arthropod that we met in Chapter 7. There are many thousands of described species of trilobites, extending from the Cambrian to the Permian – trilobites were one of the groups of animals that disappeared in the biggest mass extinction of all, the end-Permian Great Dying of 252 mya. We know that trilobites were in existence by 520 mya. And it seems likely that some of them, for example species of the order Redlichiida, were already in existence by 530 mya. But no trilobite fossils are known from 540 mya, and given their great fossilization potential, this probably means that at that early stage of the Cambrian the first trilobite – at least as we know them – had yet to evolve.

What animals pre-date the trilobites? The main indications of animal life from about 540 mya are the 'small shelly fossils' that are found at many localities across the world. These consist of the broken-up hard parts of a variety of marine invertebrates, probably including molluscs, brachiopods, sponges, echinoderms, and the mysterious shelled halkieriids. This last group has a bizarre body-form described by English palaeontologist Simon Conway Morris and his colleague John Peel in 1990; their exact taxonomic position is debated. The existence of the small shelly fossil deposits indicates that several invertebrate phyla probably existed at the start of the Cambrian.

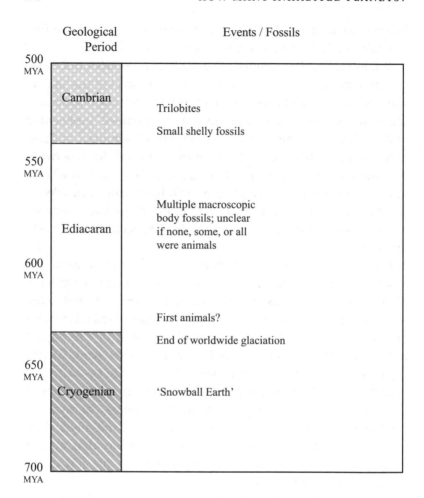

**Figure 12** Narrowing down the origin of animals. From a broad-brush starting point that the first animal lived between 0.5 and 1.0 billion years ago (an uncertainty span of 500 million years), it's possible to bracket the origin of animals to the period from 635 to 535 mya (a span of only 100 million years), based on a mixture of fossil evidence of animals in the early Cambrian and geological evidence of a worldwide glaciation lasting until the start of the Ediacaran. However, the levels of confidence that we can place on these two figures is different: after 535 mya there were *definitely* animals; before 635 mya there were *probably* no animals.

Although the small shelly fossils extend at least five million years further back into the earlier Ediacaran period (541–635 mya), the main animal-like fossils of those times are collectively referred to by the same name as the period itself – the Ediacaran fauna, or Ediacaran biota. Much rests on the choice made between these two descriptors of the organisms concerned, because with 'fauna' the implication is that they were animals, whereas with 'biota' the kingdom-level affiliation is left open. The latter option is preferable for now, given the enigmatic nature of these organisms. If at least some of them were animals, then the earliest animal may have lived about 570 mya. Recent evidence in favour of some Ediacarans being animals has come from two directions: molecular evidence reported by Ilya Bobrovskiy and colleagues in *Science* in 2018 that the already-known *Dickinsonia* belongs in the animal kingdom; and the discovery of a new Ediacaran creature called *Yilingia* that is probably an animal, reported by Zhe Chen and colleagues in *Nature* in 2019.

But in fact the earliest animal probably lived considerably earlier even than that. Comparative DNA sequence data from distantly related animals suggest that the lineages leading to the various animal phyla diverged from each other earlier than 600 mya. The first use of such data suggested *much* earlier than 600, but these studies are now regarded as flawed. Several more recent DNA-based studies suggest that the first animals originated soon after the beginning of the Ediacaran period, in other words about 630 mya.

Did animals go further back than that? Before the Ediacaran was the Cryogenian period (720–635 mya) that we met earlier, with conditions perhaps extreme enough to be describable as 'Snowball Earth'. This was probably not a period of time that was conducive to the origin of animals. And it's certainly true that there is no persuasive evidence that animals existed in the Cryogenian, so perhaps our figure of 630 mya for the first animal is about right.

## How Many Planets with Animals?

Recall that our order-of-magnitude estimate of the number of planets in the Milky Way with microbial life is 1 billion. We now ask what proportion of these would also be expected to have animal life. One way to approach this question is to use the information in the previous section as follows.

Exoplanets that currently exist in the galaxy are a variety of ages. They include those too young to have any life at all, those that may have life right now, and those orbiting dying stars that may once have had life. Restricting our attention to those that do have life, there is still a wide range of ages. Suppose for a moment that all these inhabited planets have evolutionary processes that are broadly similar to that on Earth. In this case there will be a microbes-only phase, a microbes-and-animals phase, and a 'terminal phase' that's hard to predict (because Earth hasn't reached it yet) but might well be characterized by a continuing mix or by microbes only. The latter seems more likely, given the greater occurrence of extremophiles among microbes than among animals.

Suppose the evolutionary origins of the various types of life elsewhere (microbial, animal, intelligent) are spaced approximately in the same way as on Earth. Then on the basis of Earth's history to date, the fraction of microbially inhabited planets that also have animals can be estimated by the relative durations of these two types of life here, which is 630 million compared to 4 billion years, giving a fraction of 16%. Thus we might expect to find animal life on *160 million planets*. If we consider a possible future that is symmetrical to the past, with microbes outliving animals by the same amount of time that they pre-dated them, then the calculations remain the same for total planet lifespan as they were for the first half of it.

Of course, not every inhabited planet will turn out to be even broadly similar to Earth. Some may never evolve animals at all. So the above predicted number of animal worlds is too high, though it's impossible to estimate by how much. So, again taking an orders-of-magnitude approach, let's say that we shouldn't expect the number of animal worlds to be more than 100 million, and 10 million might be a safer bet. Even 10 million is a very large number of potential animal kingdoms; but it's also a very small fraction of our guestimated trillion planets in the galaxy – 0.001%. So it's a case of Common Earth (numbers) or Rare Earth (percentages). Since everyone agrees that Earth-like biospheres will be rare in percentage terms, it's the former way of looking at things that is more important here.

The question of whether all microbial worlds eventually evolve animals is part of a larger question – how repeatable is the evolutionary process, both from time to time and from place to place? That's where we're headed next.

*Part IV*

# Discovering Life

## Key Hypotheses

### The King Carbon Hypothesis
Carbon-based life is the most probable, and hence the most common, form of life in the Milky Way, and indeed in the universe. However, it may not be the only form of life. We should keep an open mind on this; the King Carbon hypothesis should not give way to carbon chauvinism.

### The Sequence Specificity Hypothesis
Life requires a type of macromolecule that exhibits sequence specificity that is similar in general, though not necessarily in detail, to the specificity that is found in nucleic acids and proteins. Such a macromolecule need not belong to either of these groups; however, molecular length on its own, without sequence specificity, is insufficient as a basis for life.

### The Cell Centrality Hypothesis
Membrane-enclosed cellular life is the norm, but in the vastness of the galaxy – and even more so in the universe – there will turn out to be some cases of life taking a different physical form.

# 13 ON THE REPEATABILITY OF EVOLUTION

## Replaying the Tape of Life

Stephen Jay Gould was one of the leading evolutionary biologists and science popularizers of the late twentieth century. One of his many claims to fame was his 1989 book *Wonderful Life*, giving his interpretation of the Cambrian animals of the Burgess Shale fossil deposits in British Columbia. These animals were the products of the Cambrian explosion, whether we think of it as a real explosion or an explosion of fossilization potential. Gould promoted a view of these animals that emphasized their degree of morphological difference – disparity – both from each other and from the animals that came after the Cambrian. Connected with this, he emphasized the importance, in evolution, of the role of historical contingency.

Most biologists now think that Gould went too far in his emphasis on disparity. Animals that he saw as belonging to now-extinct phyla are these days seen as fitting comfortably within phyla that we recognize from the current fauna. For example, the legendary *Anomalocaris* ('strange crab'), which Gould thought belonged in a new phylum, is now thought to have been an arthropod. However, Gould's point about the role of contingency transcends his interpretation of the Burgess Shale animals and remains an interesting comment on the way evolution works.

The basic idea is this. Although natural selection is a deterministic and to some extent predictable process, its role is strongly influenced by one-off chance events like whether 'species X' survives a mass extinction event or not. If it does, natural selection will modify it in ways dictated by subsequent environmental pressures, and may in the long term produce from it a large group of descendant species that share certain characteristics. But if species X does not survive the extinction event in question, then there will be no great group of descendants. The difference between surviving and not surviving a major extinction event could be a matter of the life or death of a few individuals of the species concerned living in a particular geographic location. And yet not only are the consequences already described considerable in themselves, but they have knock-on effects further downstream in the evolutionary process. For example, the existence of a large X-derived group will set up selection pressures on other groups due to, for example, predatory interactions; while the absence of that group, and the possible presence of a large alternative group, will significantly alter the overall selective regime.

In terms of the Burgess Shale fauna, if any of the anomalocarids had left descendants that had survived beyond the Palaeozoic era, then there would have been certain selective pressures on later organisms, perhaps right up to the present day. It is impossible to assess how different the results of evolution in other animal lineages would have been, had this occurred.

Since we cannot do real experiments relating to how evolution works at this grand level, we have to take other approaches. What Gould famously did was the thought-experiment of 'replaying the tape of life'. To be specific, he imagined that we roll back the entire series of events that have occurred in evolution to some early point in time, for example part-way through the Cambrian period, and then let the natural world restart from that point. His argument was that even given the same starting point and the same environmental conditions, the array of animals that would evolve might be very different from the ones that actually did evolve in the real world as we know it.

To some extent, such a thought-experiment is just a restatement, and perhaps a refinement, of the question of what would happen if evolution were to be restarted from a certain point in time. We have no idea, and nor did Gould, what the answer would be. So we are all free to follow our own preferred model. However, one thing that can be said for certain is that the result of replaying the tape of life – were such an experiment possible – would be strongly constrained by the general features of life on planet Earth that had become well established at the point of replay. For example, evolution rebooted from the Cambrian would not invent another genetic code. Nor would it abandon cellular construction in favour of some other way of building bodies. And at the deepest level of all, it would hardly depart from being based on organic (carbon) macromolecules. Yet if the tape is taken back to its very start, instead of an intermediate point between then and now such as the Cambrian, anything is possible – or is it?

## Changing Place as Well as Time

Imagine being an observer of the early universe – say about a million years after the Big Bang. By that stage, the era of primordial nucleosynthesis in which nuclear fusion throughout the hot nascent cosmos produced helium nuclei from hydrogen was a distant memory (having happened in the first half-hour). The era of recombination, in which ions and electrons got together to produce the first neutral atoms of hydrogen and helium instead of just their nuclei, was also in the past. Star formation is still in the future, and so too are the very first planets. There is no life of any kind anywhere; so you as an imaginary observer are quite alone.

Now proceed forward in time. Eventually, due to heterogeneity in the density of matter – perhaps dark matter first followed by normal matter – collapsing dense clouds get hot enough for nuclear fusion to begin. Thus the first stars switch themselves on – and in doing so produce the first new light since the end of the primordial nucleosynthesis era. These first stars may have proto-planetary discs, but even if they do, no rocky planets or life are possible because the metallicity of the entire universe at this early stage in its history is virtually zero. There is no carbon, no oxygen, no nitrogen, no phosphorus. In other words, none of the elements that we now think of as the main building blocks of life are yet available.

But if we move further forward in time until the first ever stars are dying, we observe the formation of the heavier elements within them, and the expulsion of these elements into what is now interstellar space. In the case of the larger stars, this expulsion happens sooner (because big stars die young, as we've already seen), and is more dramatic – taking the form of supernova explosions. Eventually, the metallicity of the gas clouds that will collapse and form later stars is sufficient for these stars to have rocky planets, and for some of those rocky planets to evolve life. Exactly when in its history the universe thus had its first habitable zones in combination with life-generating elements is unclear, but it was well within its first billion years.

Every planet in the habitable zone around its star from way back then to the present day was – or is – a possible host for life. The number of planets that are now inhabited is probably huge, as we've seen. But this number is only a fraction of all the planets that have ever had life, just as the number of species of life-forms on Earth today is only a fraction of all the species that have ever existed here. If the hypothesis of Autospermia is true (and its nemesis Panspermia false) then every one of these currently or previously inhabited planets is (or was) an independent experiment in the origin and evolution of life; or, to go back to our Gouldian starting point, an independent 'tape' of life.

Now we come to the question of which event was the *playing* of the tape, as opposed to its replaying. From Gould's perspective, where possible life on Earth was the issue, the actual Earth as we see it was the playing event; all the hypothetical scenarios of thought-experiments were replayings of the tape in our minds. There is no historical sequence of replayings here because none of them are real – we can imagine a series of different possible futures simultaneously.

However, in the astrobiological equivalent of Gould's thought-experiment, there most certainly *is* a historical dimension to the playings. There must have been a planet somewhere on which the first ever life arose – though perhaps that apparently self-evident statement needs to be qualified. If the universe as a whole is infinite, then simultaneous origins on different planets must have been at least a possibility. But in the context of the Milky Way the statement is probably true enough. Either way, Earth is simply one of numerous playings. Not only are we not alone in terms of life in the galaxy, but from a probabilistic point of view we were not the first.

Let's suppose that our order-of-magnitude estimates from the previous chapter of about 1 billion planets with life and 10 million with animal life in the Milky Way are correct. Such vast numbers present much opportunity for life to evolve in very different forms on different planets. Whatever variation exists, doubtless much of it is due to different conditions prevailing on the different planetary homes, and some of it is due to historical contingency. But this is where we get to that hugely important yet also amazingly difficult question of *just how different* life elsewhere is from life here. Human thought about this issue is so constrained by what we know about life on Earth – and by what we don't yet know of life on exoplanets – that it's almost tempting to admit defeat right at the start. But that's not in human nature. Let's now boldly think our way into the unknown, while always keeping in mind the pitfalls of this process.

The rest of this chapter is structured as follows. First, we consider whether life might be possible based on some element other than carbon. Then we examine the possibility that it might be based on carbon but not on cells. Next we contemplate the likelihood of becoming multicellular, and the types of multicellular creature that might exist beyond the Earth. Finally, we imagine worlds with and without intelligence.

## King Carbon

When searching Google Images about a year ago for pictures with visually attractive artists' impressions of alien life, I came upon one with an assertion attached to the image, as follows:

'Scientists say silicon-based life-forms are as likely as carbon-based.' Personally, I think this is nonsense. Nevertheless, when people are asked to suggest an alternative element to carbon as the basis for extraterrestrial life, by far the most commonly suggested is silicon. It's worth asking why.

What features are needed in an element that might form the basis of life on one or more planets? A short answer is reactivity. The element concerned must be able to form bonds with many other elements and thus be capable of generating complex macromolecules. Indeed, the word 'complexity' is key. Although we sometimes think of 'simple' (microbial) life versus 'complex' (animal and plant) life, this version of 'simple' only works as part of a particular comparison. If instead we make the kind of comparison that is now in focus – life versus non-life – then the chemistry of microbes is complex in the extreme.

Let's think of a type of macromolecule that's quintessentially characteristic of life on Earth – DNA. A single macromolecule of this amazing stuff in a living system, whether microbial, animal, plant, fungal, or 'other', consists of thousands of sugar residues connected together lengthwise into two strands in which they alternate with phosphate groups, with the strands being connected 'widthwise' by nitrogenous bases, each of which is linked (a) to one of the sugar residues and (b) to a base on the opposite strand (by hydrogen bonds). The two strands, of course, coil around each other to form the famous double-helical shape first suggested by James Watson and Francis Crick at Cambridge in 1953.

In one DNA molecule, such as the single circular chromosome of a bacterium, there are at least 100,000 pairs of nitrogenous bases (and usually far more). Each base, and each sugar unit, has several carbons. So, even in the simplest of bacteria with the smallest of genomes, there are probably more than a million carbon atoms altogether. And of course there are also very large numbers of atoms of the elements hydrogen, nitrogen, oxygen, and phosphorus. But the crucial thing for living – and hence reproducing – systems is not just that the key macromolecules of life are very large, but also that they can reproduce or copy themselves in some way. We now know many of the details of the way in which DNA copies itself (in a 'semiconservative' manner in which the double helix unzips and a new helix is made from the template of each of the original ones). However, even before these details were known, the double-helix model suggested an inherent ability to self-copy. As Watson and Crick wrote in wonderfully understated style: 'It has not escaped our notice that the specific pairing we have postulated immediately suggests a possible copying mechanism for the genetic material.'

Now we need to ask whether some other element than carbon could provide a basis for macromolecules that can reproduce themselves in an equally predictable and faithful manner, albeit using some different mechanism than that of DNA. In particular, let's look at this possibility for silicon. The mineral quartz is a large molecule consisting of a framework of many silicon and oxygen atoms. A crystal – whether of quartz or some other mineral – can be very large. And it can also grow. To some extent this growth can be likened to reproduction because it involves additions to the initial framework or lattice, and in each new part of the crystal the structure of the lattice is the same as in the old parts – so the old parts could be said to have reproduced themselves.

However, there is something very important missing here from the perspective of life: specificity. Any one part of a quartz crystal looks pretty much like any other part. This situation is very different from a DNA molecule, where every linear stretch of a reasonable length is very different from every other one – except in the case of the various categories of repetitive DNA, which needn't concern us here. So, along the coding DNA of a typical gene, the sequence of particular nitrogenous bases of (say) the first 12-base stretch is different from the sequence of the next 12-base stretch, which in turn is different from the next, and so on (Figure 13). Life arises from this mixture of specificity and reproduction, not from reproduction alone. No particular form of specificity is crucial – for example the form embedded in the triplet-based genetic code of life on Earth – but *some* form of specificity is essential. This is what I call the Sequence Specificity hypothesis.

**First 12-base stretch:**

| | DNA | – | GAA | GGA | CTA | CCC |
|---|---|---|---|---|---|---|
| | mRNA | – | CUU | CCU | GAU | GGG |
| | protein | – | Leu | Pro | Asp | Gly |

**Second 12-base stretch:**

| | DNA | – | ACA | CAA | CTT | TAC |
|---|---|---|---|---|---|---|
| | mRNA | – | UGU | GUU | GAA | AUG |
| | protein | – | Cys | Val | Glu | Met |

**Primary structure of the (short) protein:**

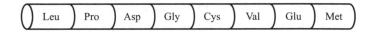

| Leu | Pro | Asp | Gly | Cys | Val | Glu | Met |

**Higher-level structure of the protein:**

**Figure 13** Example of the sequence at the start of the coding region of a particular gene (DNA), together with its corresponding message (mRNA) and the resulting protein. The first and second stretches of 12 bases are shown, each coding for four amino acids, due to the universality of the 'triplet code' in which three consecutive bases in DNA and RNA correspond to a single amino acid in the protein. The specificity of the overall eight-amino-acid sequence (compared to other such sequences) is very clear. However, a typical protein is far longer and thus even more specific, for example about 140 amino acids in one of the protein chains of a haemoglobin molecule (with the exact length depending on the species and the variant).

Are there any large silicon-based molecules that might have this combination of specificity and reproduction? I don't know of any, and I'm sure that if one was discovered it would make headlines around the world. Also, synthesizing such a molecule would be a real coup for a research group, for they would then have a basis for a form of artificial life. At present, studies in this field can be divided into those of robotic life, software life, and life based on artificially synthesized DNA. There is no network of laboratories pursuing the possibility of self-sustaining silicon-based life. The reason is probably that no-one can envisage how to go about making a silicon-based equivalent of DNA.

The distinction between 'carbon-based' and 'silicon-based' molecules is useful, especially in the context of possible chemical bases for life. However, we should not fall into the trap of thinking that molecules with these two elements are mutually exclusive – they're not. There's a large class of chemicals called organosilicon compounds (such as the silicones used in sealants), which contain both carbon and silicon; indeed, this class is defined by the presence of bonds between these two elements. It's interesting to note that on Earth there are no organosilicons in any life-form that has been studied in this respect, and therefore probably no organosilicons in any terrestrial life. Whether they exist in any extraterrestrial life we do not yet know, of course. However, they lack the sequence specificity to provide the main informational basis for such life.

Remember that we are supposed to be bearing in mind the pitfalls of thinking our way into the unknown. It's important to keep an open mind. So we should not become carbon chauvinists who make the assumption that life based on carbon is the only possible kind of life anywhere, and at any time. However, perhaps we should be wary of the opposite extreme too. What would that be? Perhaps suggesting that silicon-based life and carbon-based life are equally probable, as in that wild claim I found on the web when searching for plausible artistic aliens. Or perhaps going further still and suggesting that almost any chemical element could provide a basis for life, even the likes of helium or argon. That would, I think, be folly.

One of the reasons why silicon is more often touted as a possible alternative basis for life than other elements is that it belongs to the carbon group in the periodic table. In general, a periodic table group is characterized by similarities among its member elements in the outer shell of electrons of the atom. In the case of the carbon group in particular, the number of such electrons is four, which can be contrasted, for example, with the fluorine group (the halogens) where the outer shell has seven electrons.

Although a similar outer shell of electrons conveys some chemical similarities to the members of a group of elements, such similarities should not be overstated. As well as carbon and silicon, the carbon group (alias group 14) includes tin and lead. Few scientists have argued in favour of these elements providing possible bases for extraterrestrial life.

So, our conclusion in the present section should be as follows. We should avoid carbon chauvinism, which could be described as a stance or a mind-set. We should acknowledge that in a vast galaxy life could take many forms, including forms that are hard for us to envisage. However, from a pragmatic perspective of gearing up to search intensively for life in the Milky Way, or at least our own little neighbourhood of it, we should focus our search on carbon-based life. The reason for this approach is that at least we know carbon-based life is possible. It may also be the most probable, and hence the most common. I call this conjecture the King Carbon hypothesis.

## The Centrality of the Cell

All life on Earth is carbon-based. In contrast, although most of it is based on cells, some of it is not. This seems to be telling us that cells are not necessary for life on Earth, but paradoxically this may not be true. To see why not, we need to look at two things: (1) what features define a cell, and (2) the nature of life cycles.

Here's a reasonable working definition of a cell. It is a unit composed largely of quasi-liquid cytoplasm that is bounded by an integral external membrane and contains either a roughly defined nucleoid (prokaryote cell) or a clearly defined nucleus (eukaryote cell), wherein the majority of its genetic material can be found. To illustrate how difficult it is to come up with a *precise* definition, this working one would exclude the mammalian red blood cell, because, in its mature state, it lacks a nucleus. This could be dealt with by including in the definition some reference to having a nucleus in its progenitor cell. However, there are other complications as well as this one, and elaborating the definition to cope with all of them would be cumbersome. Anyhow, for our purposes here a working definition will suffice.

Here's a typical animal life cycle: germ cells → fertilized egg cell → early embryo with few cells → later developmental stages with many cells → reproductively mature adult → germ cells. Again, there are some ifs and buts. Not all life cycles involve sexual reproduction, so 'germ cells' (eggs and sperm) will not always be applicable. Many plants have 'alternation of generations' consisting of haploid and diploid versions of the same plant, with very different body-forms, as is the case with ferns. Some types of organism (though no types of animal) remain unicellular throughout their life cycle, and so on. But again a rough idea is all we need.

There is a group to which most of the above generalized life cycle does not apply: the acellular slime moulds. We met these briefly in Chapter 3 – there are about 1000 species of them worldwide. Despite lacking cells, the adult can reach a breadth of up to about a metre. The life cycle has a unicellular stage, but what develops from that stage is a large bag of cytoplasm with many nuclei but no internal cell membranes. These odd organisms are found in many different habitats. They are often found growing on decaying wood on forest floors. Often they are colourful – some species are bright yellow.

The growth of an acellular slime mould involves nuclear division, which happens in much the same way as in an animal or plant embryo. However, this division is not accompanied by subdivision of the cytoplasm. So, while the amount of cytoplasm increases, and the outer membrane extends so that it continues to encompass the enlarging body, the lack of internal membranes means that the organism can end up with a million cohabiting nuclei. Naturally, this great bag with many repeats of the genome inside it has no supporting macro-skeletal structures. So, unlike large animals and plants, it can never grow upward against the force of gravity. Thus although slime moulds can reach impressive surface areas, they are always flattish.

To a biologist it is clear that these creatures are non-cellular. Their name, incorporating the adjective 'acellular', is based on their perceived lack of cells. And yet in two important senses this name is inappropriate. First, they all go through a unicellular stage in their life cycle, so as four-dimensional organisms they are not truly acellular. Second, to a materials scientist the difference in construction between them and a multicellular equivalent – say a large fungal mould – might not seem very pronounced, especially in relation to all those other modes of construction that are used in the human world. Looked at in this way, an acellular slime mould is just a minor variant on an organism from its sister group – a cellular slime mould. In both cases, there is an outer lipoprotein membrane, a lot of internal cytoplasm, and many copies of the genetic material. The only difference is the lack of internal lipoprotein membranes separating one local blob of cytoplasm from another in one of the two groups.

The other organisms that we should consider here are the viruses. These might seem better candidates for being non-cellular life-forms than are slime moulds. A typical virus consists merely of a nucleic acid genome packaged in a protein coat. In terms of size, a virus is several orders of magnitude smaller than a cell. There is no cytoplasm; there are no membranes; there is no metabolism. Because of the last point, viruses are often considered not to be life-forms at all. They do reproduce, but only when they have access to a cell of their host organism – which can be anything from a bacterium to a human. So, to the question 'are viruses non-cellular life-forms?' we can answer 'no' either because they are not life-forms at all or because they cannot complete a life cycle without a cell, albeit one belonging to another species.

We end up concluding that no life on Earth is acellular throughout its life cycle, and most forms of life here are cellular throughout their entire life cycle – or at least the vast majority of it. This final qualification is necessary because much more familiar organisms than slime moulds and viruses have acellular stages in their life cycles. Insects have an early developmental stage in which the nuclei have divided many times since the fertilized egg stage but the cytoplasm has not, with the result that the embryo is effectively a bag of a few thousand nuclei. And part of the mammalian placenta also takes the form of a large multinucleated sac.

Given the overwhelming predominance of cells, and the universality of lipoprotein membranes in all metabolizing life on Earth, should we expect such membranes to characterize life elsewhere? We have to be careful to avoid a tautology here. If we use possession of a membrane as one of the features that define life, the question becomes meaningless. But suppose we restrict the definition of life-forms to the two key features of metabolism and reproduction, what then? Here's a hypothetical answer to this question – let's call it the Cell Centrality hypothesis. Membrane-enclosed cellular life is the norm, while in the vastness of the galaxy – and even more so in the universe – there will turn out to be some cases of life taking a different physical form.

## Form and Function

Suppose that there are many biospheres spread across the Milky Way in which the life-forms are both carbon-based and cellular. And suppose that in those biospheres that last long enough – for example those whose host planets orbit Sun-like or smaller stars – large multicellular life-forms eventually evolve from small unicellular ancestors. What are those large life-forms like? The literature of science fiction is overflowing with possibilities, many of which are clearly based on animals (and, though less often, plants) that can be found on Earth. Is this syndrome of the 'strangely familiar extraterrestrial' merely a result of the limitation of humans' ability to go far beyond the known? Or is there a chance that most biospheres we eventually discover will be populated by creatures that we recognize as animals or plants, perhaps even (say) as insects or flowers?

Before going further into this issue, we should distinguish similarity in form from similarity in function. Biologists often divide the study of form – anatomy, morphology – from the study of function – physiology – while recognizing that there is a link between the two. But the link is far from rigid. From an ecological perspective – another aspect of function – an animal can be a herbivore of savannah ground flora while having the structure of a small insect or a large vertebrate. A plant can be an inhabitant of boggy environments because it has no water transport system – mosses, for example – or because it likes the acid soils that prevail in most boggy areas – as in the case of the flowering plant known as bog cotton.

Closely related species can have different functions but rather similar forms. Within the great apes, chimps, bonobos, and humans all eat vertebrate prey as well as other foods, while gorillas and orangs do not. Most species of centipedes, including members of the genus *Strigamia*, live in inland habitats, but one species of that genus lives only at the coast. The difference in form between this species and its closest inland relative is minimal, yet one has a diet of intertidal invertebrates, the other a diet of invertebrates that are found in the interstices of the soil. With both the connection and the distinction between form and function in mind, let's look briefly at two issues in relation to the evolution of multicellular organisms on exoplanets.

First, function, and in particular the fixation of carbon: recall from Chapter 5 that this takes two forms, namely chemosynthesis and photosynthesis. Here on Earth, the balance of carbon fixed – and energy produced – by the two processes is strongly biased in favour of the latter. And a reasonable expectation for a biosphere at the surface of an exoplanet is that photosynthesis will also be the predominant form of fixation there, simply because of the abundance of sunlight. There might, of course, be exceptions. A dense atmosphere, like that on Venus, can drastically reduce the amount of light reaching the surface. If there's life in the sub-surface oceans of Europa or Enceladus, under a thick sheet of ice, then it might be based on chemical sources of energy rather than on light. This would also be true of rogue-planet biospheres, if any of these exist. However, exceptions aside, we should expect photosynthesis to prevail.

But recall that there are two types of photosynthesis – oxygenic (water being split and oxygen produced) and anoxygenic (e.g. hydrogen sulphide being split and elemental sulphur produced). Again, the balance of energy generated by the two on present-day Earth is highly asymmetric, with oxygenic photosynthesis being the norm. In an extraterrestrial biosphere with a long evolutionary history and plenty of water, we should perhaps expect the same to be true. This expectation is very important in relation to searching for life by analysing exoplanet atmospheres. Although searching for an atmosphere with an appreciable concentration of oxygen (and some ozone) seems like a good strategy for astrobiologists, we need to keep in mind the possibilities of (a) abiotic production of oxygen and (b) biotic forms of carbon fixation that don't produce oxygen.

Now we turn from function to form, and from plants to animals. What forms might extraterrestrial animals take? By definition an animal is multicellular and heterotrophic. At first sight this is so broad a definition that it seems reasonable to expect to find animals elsewhere that look nothing like those on Earth. But perhaps ecology narrows things down somewhat. Most animals on Earth are mobile; and in general mobility helps if you have to locate food. Mobility imposes certain restrictions on body-form. And the type of motion is important too, along with the interconnected issue of the medium in/on which movement occurs. If the Parallel Places hypothesis of Chapter 6 is correct, then extraterrestrial animals may move on land, in water, and through the air, just as they do on Earth. If so, parallel evolutionary radiations might be expected. They're probably not quite as similar to those on Earth as some of ours are to each other – such as the radiations of marsupial and placental mammals we discussed in Chapter 3. Nevertheless, some broad architectural features are to be expected, including skeletons.

## Elusive Intelligence

While many lineages on Earth are characterized by skeletons, only a few are characterized by intelligence. As we saw in

Chapter 8, the exact proportion depends on how intelligence is defined. Crows pass the sequential tool-use test, dolphins pass the mirror test, and octopuses pass the bottle-opening test. Chimps pass them all. But in terms of looking for intelligent life elsewhere in the galaxy, these levels of intelligence are unimportant for pragmatic reasons: none of them help us *to search for* intelligence. We can only search for intelligence, as opposed to searching for life in general, if we use techniques that are related to features of technological civilizations – and in particular the sending and receipt of radio waves.

One way to look at this issue is to contemplate distant aliens trying to find signs of life on Earth. They might look for evidence of oxygen in our atmosphere (a sign of photosynthesizing life) or they might look for radio signals (a sign of intelligent, techno-capable life). Such searches of Earth might have found evidence for photosynthesizing life a billion years ago. But searches for radio messages would only have become successful about a century ago, because only a single lineage here has evolved to be able to send and receive radio signals, and that lineage has only had this capability since the work of Guglielmo Marconi and others, around 1900.

Now turning around from contemplation of aliens looking for life on Earth to ourselves looking for intelligent life elsewhere, we come up against a serious problem. Any quest in astrobiology suffers from having a starting point that is a sample size of one – a single known life-bearing planet. But at least for many specific quests we have the advantage of multiple lineages. For example, the number of lineages with oxygenic photosynthesis compared to the number with its anoxygenic equivalent may be useful in helping us to think about the likely relative commonness of the two types of photosynthetic life on another planet. But when it comes to advanced intelligence, we are in an even worse predicament: not only do we only know of one planet with intelligent life, but we only know of one lineage (out of millions) on this planet that has radio-level intelligence.

So, the likelihood of evolution on exoplanets eventually producing intelligence is one of the greatest unknowns, in a field that already has far too many of them. The following few points are probably useful guidelines in thinking about this matter – but they're no more than that. First, evolution on an exoplanet may never produce intelligence, even if it produces animals. Second, in those cases in which intelligence *does* result, the route to it will probably be very long, so we should not expect to find intelligent life on young planets. Third, intelligent life might have a tendency to self-eliminate within a few centuries of its inception. Fourth, notwithstanding the above three points, there may be intelligences in our galaxy and beyond that make our own level of this trait look feeble. That's enough for now; we'll return to intelligent life in Chapters 16, 17, and 20.

# 14 CANDIDATE PLANETS

## Managing Expectations

Having now got a rough idea of what we're looking for in terms of extraterrestrial life, let's see where we might find it among the exoplanets discovered thus far. In this chapter, we'll look at a few promising planets individually. However, it's useful to do some quick general calculations first, based on the guestimates of Chapter 3 (number of planets in the Milky Way) and Chapter 12 (fraction of these inhabited by various types of life). Although these are only orders-of-magnitude figures, they're better than no figures at all.

So, our starting points are: one trillion planets in the galaxy; one in 1000 of these (1 billion) inhabited by microbial life-forms; and one in 100 of these microbial worlds having complex life in the form of animals (10 million). Thus for every 1000 planets we would expect just one to have microbial life and typically none to have animal life. Since the current tally of confirmed exoplanets at the time of writing is just over 4000, we might expect that about four would have microbes, whereas the likelihood is that there will be none with animals – or perhaps one, if we're very lucky. There will almost certainly be none with intelligent life.

The number of known exoplanets will rise over the next few years, partly because of further analyses of data acquired from Kepler, and partly because of discoveries made using new space telescopes, notably TESS (the Transiting Exoplanet Survey Satellite), which was launched in April 2018, CHEOPS (the Characterizing ExOPlanets Satellite), launched in December 2019, and the much-delayed James Webb Space Telescope, which perhaps will finally launch in 2021. Further exoplanet discoveries will follow in the 2030s after the launch of next-generation direct-imaging space telescopes that will be able to analyse their atmospheres; these will be discussed in the following chapter. However, even if the number of known exoplanets rises dramatically, the expectation would be that the number of microbial worlds will still be small, and the number of animal kingdoms proportionally smaller. For example, when our current total of 4000 exoplanets rises to 10,000, we would still only expect to have about 10 microbial worlds and our tally of worlds with animals would still be either 0 or 1.

What these figures suggest is that we need to choose very carefully which planets to discuss here, because a careless choice would almost certainly lead us to examine a series of barren planets – a barrenness that might well be confirmed by future atmospheric analyses. We're going to look at just four planetary systems now – one in each of the following four sections – so I've tried to choose them on the basis that all four look promising at first, and with luck at least one will still look promising by the end of the chapter.

What factors should be taken into account in choosing promising planets in terms of possible extraterrestrial life? First, we should ignore gas and ice giants and focus on rocky planets. Second, we should include only those rocky planets that are within the habitable zone, and if possible not too close to its inner or outer edge, given that there is always some uncertainty in exactly where these edges are. Third, we should ideally avoid tidally locked planets where instead of planet-wide biospheres there might only be more restricted bio-regions at best – though as we'll see, this avoidance is difficult in practice. Fourth, it would be sensible to focus on the closest planets, because these will be easier to obtain more data from in the future than those that are further away. The majority of the exoplanets discovered so far are within 3000 light years of Earth, and almost all of them are within 6000, though there are a few outliers at distances that extend up to almost 30,000. Searches for life in the near future would do well to avoid those distant outliers unless there is some very pressing reason to include them; a 3000-light-year radius is challenging enough. And although there are now some atmospheric studies on exoplanets that are about 3000 light years away, for example a report of the probable presence of clouds on Kepler-7 b by Brice-Olivier Demory and colleagues in 2013, for detailed characterization of atmospheres and other features of potentially habitable planets we should restrict our attention to a *much* smaller radius of space – probably about 100 light years.

Now we come to the nature of the star that the planets orbit. Given the short lifespan of planets orbiting large stars, and especially given the contrast between such lifespans and the lengthy timescale of evolution, we should probably ignore all planets that orbit stars of spectral classes O, B, and A; we should concentrate instead on planets orbiting stars of classes F, G, K, and M. This doesn't seem like much of a restriction, given that the former group makes up less than 1% of all stars, and thus probably also less than 1% of all planetary systems. However, of the classes F to M, M is both the commonest and the most questionable in terms of hosting biospheres, due to the phenomenon of tidal locking, which we've already discussed, and various other phenomena, including the high degree of activity (for example, stellar flares) that often characterizes such stars. This uncertainty over the potential habitability of planets orbiting M-class stars is a major problem for all attempts to predict the overall commonness of life.

As noted in Chapter 12, the most successful exoplanet hunting methods to date are the transit and radial velocity methods. Not only do these give clear evidence for the existence of an exoplanet, but at the same time they reveal the length of its year. We now know of planets whose years are just a few Earth days, or even, in some extreme cases, just a fraction of an Earth day. At the other end of the spectrum, we know of planets whose orbital periods last for many thousands of Earth years. It's hard to imagine what effects such extremely long or short years would have on life. Perhaps they don't matter, because evolution on any particular planet simply works with whatever duration of year, and of seasons, it finds there; natural selection on one planet does not take into account differences between that planet and another one. Perhaps the second most important thing connected with exoplanet orbits, after their mean distance out from the star, is not how long the orbit takes, but rather that it's not too eccentric. Very eccentric orbits, as we already noted, might lead to a planet being outside the habitable zone of its star for part of each year, even though it is within the zone for the rest of the year.

In the next four sections of this chapter, we'll look at a list of planetary features in each of four individual case studies. This information has been taken from several sources, the most important of which are the NASA *Exoplanet Archive* and the *Extrasolar Planets Encyclopaedia*. I have tried to simplify the information; most figures have been rounded. Where no precise information can yet be given, I have inserted 'uncertain'. In many cases, the form of measurement used incorporates explicit comparison with Earth (e.g. a radius of 1.2 × Earth). In other cases, the comparison is implicit (e.g. Earth is clearly within the habitable zone of the Sun). In one case – orbital eccentricity – there is no explicit or implicit Earth-related information in the list; so, for comparative purposes, here is Earth's orbital eccentricity: 0.017 (with 0.000 being a perfect circle). Now let's look at those four planets/systems, starting with – appropriately – the first to be discovered.

## The First: Kepler-186 f

After more than two decades of exoplanet discovery in which most of the worlds that were found were hot Jupiters and various other giant planets, the finding of the first small rocky planet in a habitable zone in 2014 was a milestone in exoplanetology and perhaps also in astrobiology. The finding was announced in the journal *Science* by Elisa Quintana *et al.* – their paper was entitled 'An Earth-sized planet in the habitable zone of a cool star'.

**Name: Kepler-186 f**
Discovery year: 2014 (method: transit)
Radius: about 1.2 × Earth (estimates range from about 1.0 to 1.4)
Habitable zone? Yes, near its outer edge
Tidally locked? Uncertain, but quite probable
Distance: about 570 light years
Star: red dwarf (class M1)
Length of year: 130 Earth days
Orbital eccentricity: about 0.05
Number of known planets in system: 5

This combination of features makes Kepler-186 f look promising as a possible host for life. The fact that its host star is a red dwarf means that both the star and its planetary system should be very long-lived. Its current age has been estimated to be rather similar to that of our own system, so there has been time for an evolutionary process operating at a similar rate to that on Earth to produce not only microbial life but also animal and plant life, and possibly even intelligent life. However, on the down side, we've already noted that there is considerable uncertainty about the habitability of planets orbiting red dwarfs in general, most notably because of the common problem of tidal locking.

At more than 500 light years distance from us, Kepler-186 f is not ideally placed for future studies on its atmosphere. This is not only a problem in terms of being able to detect gases – or combinations of gases – that might be considered as biosignatures; it is also a problem in terms of getting a good idea of the planet's surface temperature. We've already noted that Venus's maximum surface temperature is hotter than Mercury's, due to its possession of a thick atmosphere that includes greenhouse gases. At present we have no idea whether the atmosphere of Kepler-186 f is thick or thin; nor have we any knowledge of its composition.

The conclusion about this planet is that it has a reasonably promising combination of features for life; it has had a long enough history for life to have evolved; we don't (of course) currently know if it is inhabited; and the prospects for finding out if it is inhabited in the near future could perhaps be described as moderate at best.

## The Closest: Proxima b

The discovery of a planet in the habitable zone of our nearest star, Proxima Centauri, was very exciting to everyone involved in astrobiology. It was reported in the journal *Nature* by Guillem Anglada-Escudé *et al.* in August 2016. The discovery method was radial velocity. Since we've only encountered this method once so far (very briefly in Chapter 12), let's take a closer look at it here.

The radial velocity method is based on the Doppler effect – the alteration of wavelengths due to the movement of their source. We're more familiar with this effect in everyday life with sound waves than with light waves. A car moving at speed will seem to make a lower-pitched noise after it passes you than it did when it was approaching you. This is because the sound waves are 'squeezed up' in front of the car and 'drawn out' behind it. Squeezing up causes higher frequency and shorter wavelength, whereas drawing out does the opposite.

The same applies to the electromagnetic radiation emitted by a star. If the star is moving towards us, its radiation is shifted towards shorter wavelengths (blueshift), while if it's moving away from us, its radiation is shifted towards longer wavelengths (redshift). So, if we see a star that seems to be alternating, in relative terms, between blueshift and redshift, this suggests that something is orbiting it – probably something big, in other words a planet – and that this orbit is causing a back-and-forth 'wobble' in the star itself. In fact, both the star and the planet are orbiting their joint centre of gravity. Of course, the situation is made more complex when the plane of the planet's orbit is not pointing directly at us – but this problem can be corrected for except when its orbit is broadside-on to us (or nearly so), in which case we can't detect the wobble at all. In the early days of exoplanet hunting, this radial velocity method was the most successful. Then later, especially as we began to get data from Kepler in 2009, it was overtaken by the transit method. The two methods differ in the way they can estimate planetary size: we can get estimates of diameters from the transit method, estimates of mass (initially in the form of 'minimum mass') from the velocity method.

Now let's look at the same list of features for Proxima b as we did for Kepler-186 f, and see how they compare:

**Name: Proxima b**
Discovery year: 2016 (method: radial velocity)
Radius: roughly similar to Earth (minimum mass is
    $1.25 \times$ Earth)
Habitable zone? Yes
Tidally locked? Uncertain, but probable
Distance: 4.2 light years
Star: red dwarf (class M5.5); part of triple-star system
Length of year: 11 Earth days
Orbital eccentricity: upper bound estimated to be about 0.35
Number of known planets in system: 1

Again, this list of features suggests a reasonably promising planet for life. However, as we saw in Chapter 11, Proxima Centauri is a flare star. This means that from time to time it is characterized by temporary but major increases in brightness. These increases are thought to be analogous to solar flares, but much more pronounced, especially in relation to the size of the star. When a flare occurs, radiation of all wavelengths increases in amount; this includes damaging short-wave radiation, such as x-rays. There is not yet agreement about the degree to which such flaring is problematic for life. For this and other reasons – including probable tidal locking – we are not yet sure if the planet Proxima b is habitable. However, compared to Kepler-186 f, Proxima b is wonderfully close to us. There is even a plan – albeit a rather tentative one – to send mini-spacecraft to the Alpha Centauri system, travelling at 20% of the speed of light. This plan – Breakthrough Starshot – is part of the broader Breakthrough Initiatives, which we'll look at in Chapter 17.

## The Most: TRAPPIST-1

In 2017, a paper in *Nature* by the Belgian astronomer Michaël Gillon and his colleagues reported the discovery of three new planets in the TRAPPIST-1 system, bringing the total to seven.

Although this is not the greatest number of exoplanets known in a single system, it is unique in that all seven are small rocky planets, not gas giants, and in that at least three of them appear to be in the habitable zone.

A digression may be useful at this point into exoplanet names. There are several naming systems in widespread use. One is to name the planet after its host star – as in the case of Proxima b. Another is to name it after the telescope used to find it – as in the case of Kepler-186 f. TRAPPIST-1 is named under a variation of that second convention – it's named by an acronym (or back-ronym, in relation to Trappist monks and beer) that jointly represents the two telescopes that were used in its discovery (one in Chile and one in Morocco). It stands for TRAnsiting Planets and PlanetesImals Small Telescope. TRAPPIST-1 refers to the whole system. As usual, the star is deemed to be object A in the system, while the planets are lettered from b upward – in this case b to h. Three of these – e, f, and g – are thought to be in the habitable zone. Given the existence of three habitable-zone planets, and the fact that they don't vary too much in their features, the list that follows gives rough averages for the three.

**Name: TRAPPIST-1**
Discovery year: 2017 (method: transit)
Radius: the average for the three is roughly similar to that
    of Earth
Habitable zone? Yes
Tidally locked? Uncertain; at least in orbital resonance
Distance: 40 light years
Star: red dwarf (class M8)
Length of year: 6–12 Earth days (6 for planet e, 9 for f, 12 for g)
Orbital eccentricity: upper bound estimated to be less than 0.1
    for all three
Number of known planets in system: 7

One of the most interesting features of this system is its small size (Figure 14). Note that the star at its centre is a red dwarf of type M8. Up to this chapter, we've only looked at the main classes of stars (initial letter), not their subclasses (the following

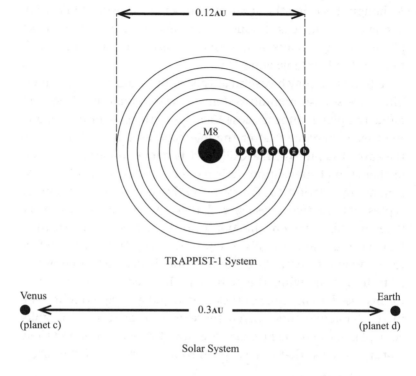

TRAPPIST-1 System

**Figure 14** The scale of the TRAPPIST-1 planetary system (top), with the distance from Venus to Earth shown (bottom) for comparison. At the centre of the TRAPPIST-1 system is a class-M8 red dwarf star; this is orbited by seven planets (b to h). The furthest-out planet (h) orbits at a distance of about 0.06 AU, so the diameter of the system is about double that figure (0.12 AU). The spacing of the planets is shown as equal here, though this is a simplification (see text). Planets e, f, and g are thought to be within the habitable zone.

number). So we're aware that the hottest stars are of class O, the coolest ones of class M. Now what we need to add to this information is the following: within any one class, the numbers 0 to 9 are used to divide it up, with 0 being the hottest within the class, 9 the coolest. Thus the M8 central star of the TRAPPIST-1 system is almost as cool as a 'proper' star can be, in contrast to the red dwarf stars of the previous two systems, which are hotter.

Because of this, the habitable zone around this star is very close in. Recall that for the solar system, and for other planetary systems too, we measure distances with astronomical units (AU). The Earth's average distance from the Sun is, by definition, 1 AU. The closest planet to the Sun, Mercury, is on average about 0.4 AU (with quite some variation around this figure, as Mercury's orbit is rather eccentric); the furthest out, Neptune, orbits at about 30 AU. In contrast, in the TRAPPIST-1 system, the closest planet (b) orbits at a distance of 0.01 AU, and the furthest out (h) at about 0.06 AU. The intermediate orbits are at approximately 0.015, 0.021, 0.028, 0.037, and 0.045 AU.

There are several possible effects of this smallness of scale of the system. One that is of particular interest here is the possibility of transfer of life (if it exists) between one planet and another. Given that the inter-orbital distances in the TRAPPIST-1 system are typically less than 0.01 AU, in other words less than 1% of the distance from the Earth to the Sun, might hardy dormant spores be able to withstand a journey in space from one planet to another? The only hard information we have that can be related to this question is the ability of some terrestrial creatures to survive orbiting the Earth when exposed to the vacuum and radiation of space. As we saw in Chapter 1, some tardigrades achieved this feat. However, it pays to probe into this headline-grabbing fact, as follows. First, most of the tardigrades died – survival was the exception rather than the rule. Second, the spacecraft concerned (one of the Russian FOTON series) was in low Earth orbit, flying at about 1500 kilometres (less than 1000 miles) above the Earth's surface. This is a mere one-thousandth of the distance from one TRAPPIST-1 planet to another. Third, the space journey taken by these tardigrades – and also by some lichens – lasted for about two weeks. This is a tiny fraction of the time that would be required to float across space from, say, planet e to planet f in the TRAPPIST-1 system. But the biggest problem for a Panspermia-type hypothesis of transfer of life-forms from one planet to another in this system is the same as it is in all such systems – how does a dormant animal or plant, a hardy microbial spore, or any other life-form achieve escape velocity? There is no FOTON rocket available to do this for them. And they cannot do it themselves.

Perhaps the most likely means of achieving lift-off is meteorite impact. Here on Earth, we receive very occasional meteorites from Mars. We think that these are blasted off the Martian surface by the impact of much larger incoming meteorites, probably from the asteroid belt. The same sort of process could serve to blast material off (say) TRAPPIST-1 planet g and send it to planet f, assuming that this system has numerous asteroids, which we don't yet know. But this means that a life-form on planet g that is potentially going to be blasted across a million miles of space to planet f has to survive an explosion first. Such a method of life transfer seems rather improbable.

## Orbiting a Sun-Like Star: Kepler-452b

Now we come to the exoplanet – Kepler-452 b – that has often been dubbed Earth 2.0, though I'm not sure this description is warranted. Some other exoplanets are closer to Earth in terms of their size, for example Kepler-186 f. To a large extent, the similarity between the Kepler-452 system and our own lies in the similarity of the suns. As we've already seen, our Sun belongs to spectral class G. Within this, it belongs to subclass G2, and has a surface temperature of about 5770 kelvins. The sun at the heart of the Kepler-452 system is also of type G2, and its surface temperature is thought to be about 5750 kelvins. (It's important to note that 'surface' for a sun/star means the top of its photosphere.) The planet Kepler-452 b has a radius of about 1.5 times that of the Earth. This means that its volume is more than three times that of Earth and, assuming it's a rocky planet, its mass is likewise at least three times the Earth's mass. So it's significantly bigger than Earth, and has a higher gravitational pull. A planet of this type is sometimes called a 'super-Earth'; indeed, the 2015 paper announcing the discovery of Kepler-452 b by Jon Jenkins and colleagues used this term. However, for the most part I'm going to avoid the term, because it has no clear definition. How super is super?

**Name: Kepler-452 b**
Discovery year: 2015 (method: transit)
Radius: about 1.5 × Earth
Habitable zone? Yes
Tidally locked? Probably not
Distance: 1800 light years
Star: Sun-like (class G2)
Length of year: 385 Earth days
Orbital eccentricity: about 0.03–0.04
Number of known planets in system: 1

Several points emerge from this list of features. First, at about 1800 light years, the system is too far away to hold out much hope for detailed atmospheric analyses in the near future. Second, the length of a year on Kepler-452 b is remarkably similar to that of a year on Earth. Third, as yet we don't know of any other planets in this system, so 452 b may be on its own. However, it would be premature to draw a firm conclusion of this kind. There may be as-yet undetected planets in the system, especially small ones and/or ones whose orbits are significantly inclined compared to 452 b such that transits cannot be observed. (Although we think of planetary systems as being 'flat', they are rarely if ever perfectly so.)

The star Kepler-452 is thought to be about six billion years old. If this is correct, then planet 452 b is probably about six billion years old too. If it has always been in its present orbit, enough time has elapsed for an Earth-style evolutionary process to have produced not just microbial and animal life but intelligent life too. Indeed, it might be home to beings far more advanced than humans. Accordingly, this planet has been a focus of attention by the SETI Institute. It has been scanned for radio signals across a wide range of frequencies for an extended period of time, using the Allen Telescope Array in Shasta County in northern California. So far, no radio signals have been detected. This may (or may not) mean that there is no intelligent life there, but it says nothing about the possibility of microbial or animal-type life. Of the four planets looked at above, I would guess that this is the most likely to host a biosphere, partly because of its G-class sun. However, its distance is problematic for future studies.

## Some Thoughts on Distance

The four planets discussed above span a distance range of 4.2 to about 1800 light years from Earth. The most distant potentially habitable exoplanet discovered so far is Kepler-1638 b, at about 2900 light years. The most distant exoplanets of any kind found within the Milky Way to date are SWEEPS 04 and 11 at about 28,000 light years (SWEEPS stands for Sagittarius Window Eclipsing Extrasolar Planet Search, a 2006 survey project that made use of the Hubble Space Telescope).

Let's start by restricting our attention to a radius of 3000 light years – though as will become clear shortly we'll need to reduce this figure. Not only does this radius include Kepler-1638 b, but it also includes the majority of exoplanets discovered so far. Now let's think back to the structure of the Milky Way, as outlined in Chapter 2. Recall that our solar system is located within the Orion arm, which has dimensions of about 10,000 × 4000 light years. So, most if not all of the habitable exoplanets discovered so far are within this region. And up to now there has been a bias towards the 'inward-pointing' direction of the arm (the direction that leads to the galactic centre), because that's where the Kepler field of view was, as opposed to the 'outward-pointing' direction (the direction that leads to the Milky Way's periphery).

Kepler's follow-up act, TESS, has a much broader field of view, as described by astrophysicist Joshua Winn, in an article in *Scientific American* in 2018. It has a very different construction – four small telescopes instead of one large one. And TESS is able to rotate in order to peer in different directions. With the combination of these two features, it should be able to see about 90% of the sky. Kepler was only able to look at a fraction of 1%. Now, with TESS up there, instead of a finger of information pointing in one direction we'll soon have a quasi-sphere of information pointing in nearly all directions.

What does this mean for the number of known exoplanets in the near future? After the TESS data have all been collected and analysed, we will know of many more habitable-zone planets than the current number of about 20. If we use the comparative fields of view as a guide, the number of such planets might rise to about 2000 (and the total number of exoplanets from 4000 to 400,000), though I suspect these numbers are overly optimistic.

In terms of distance, no exoplanet is likely to be discovered that is nearer than Proxima b, simply because there are no stars closer than its host star, Proxima Centauri. Although there is always the possibility of a 'lone' or 'rogue' planet in the region of interstellar space between here and there, such planets are unlikely to be able to host life, as we noted earlier, given their lack of a nearby source of light. It's impossible to be equally precise about what will happen in terms of exoplanet discovery at the other end of the distance scale. Assuming that our region of the galaxy isn't special, there are likely to be exoplanets throughout its vast expanse. The furthest from us are probably about 80,000 light years away. And of course we don't think our galaxy is special either. So there will be exoplanets scattered through Andromeda and galaxies further afield.

But the key to making progress right now is to concentrate on the closer end of the distance spectrum, not its as-yet unknowable outer counterpart. How close we need to look depends on how we are looking. A limit of about 3000 light years (which is where we started this section) is broadly appropriate for some types of studies. However, for future investigations of the detailed composition of the atmospheres of non-transiting exoplanets using direct-imaging space telescopes (such as NASA's planned HabEx and LUVOIR – see next chapter), we will need to restrict our attention within a much closer distance limit. As noted earlier, this limit would be about 100 light years.

# 15 ATMOSPHERIC SIGNATURES

## Signs of Life

When investigating another world, whether a local one such as Mars or a distant one such as any of the four exoplanets discussed in the previous chapter, the key question for an astrobiologist is: are there any signs of life? This then raises another question: what exactly should we take as signs of life? And we also need to distinguish signs of present life from signs of past life. These questions take us into the realm of biomarkers and biosignatures.

Care is needed here because these terms are inconsistently used in astrobiology, and 'biomarker' also has a diversity of usages in various other branches of science. In cases where the two terms are used by the same author, they are sometimes treated as synonymous, sometimes as distinct. So the best thing I can do here is to adopt a policy, explain it, and stick to it. I will avoid the term 'biomarker' entirely, because of its diverse usage, and I will use only 'biosignature'. The definition of a biosignature will be 'any information that could be reasonably regarded as evidence for present or past life'.

Such life-suggesting information can come in a variety of forms: molecular, morphological, spectroscopic, and a catch-all category of 'other'. Some forms may be relevant just to past life or just to its present-day equivalent; others may be relevant to both. If one day we find a fossil of fish-like form on Mars, that would be strong evidence of past Martian life, provided there had been no skulduggery by the scientists involved. Secretly transporting a fossil to Mars and placing it where it will be discovered by others would make the notorious Piltdown Man hoax pale into insignificance. Fossil microbes would be less strong evidence of past life, because the amount of morphology visible on one of these is much less than on a fish, and so the likelihood of confusing a biosignature with some form of abiogenic chemical signature is much higher. Indeed, we saw an example of this confusion in Chapter 10, where supposed fossil microbes on a Martian meteorite turned out to be explicable in solely chemical terms.

The macromolecules of Earth-based life, such as DNA and proteins, can be regarded as biosignatures. However, it's important to stress that their component small molecules, such as sugars and amino acids, cannot. We've already seen that simple organic molecules are found in space and on meteorites. Their formation does not require the enzymes of life. But connecting up chains of hundreds or thousands of them does require enzymatic activity. Thus if a meteorite arrived on Earth tomorrow and was found to contain an amino acid – say glutamine – that would not be big news. However, if it was found to contain a protein – say haemoglobin – that would be sensational, as we noted in Chapter 1.

Virtually all meteorites that arrive on Earth have originated in the solar system – usually from asteroids, occasionally from other sources such as the Moon or Mars. The first large rocky body observed to arrive in our solar system from elsewhere was 'Oumuamua, which we met briefly in Chapter 9. Of course, there may have been earlier unknown transient visitors, especially smaller ones, which are harder to detect. However, all 'solid information' (in the literal sense) that we have about bodies beyond Earth is local; this includes samples of moon rock obtained by astronauts on the various Apollo lunar landing missions.

Since we cannot visit exoplanets, and since we do not have any meteoritic data from any of them, we must look to other types of biosignature than solid ones in our quest for exoplanetary life. And the obvious candidate is the type of signature that can be called atmospheric or spectroscopic. This is where the atmosphere of an exoplanet leaves an imprint on the light that reaches us from its system. In some cases, this imprint may be a biosignature; in other words, it may point to the likely existence of life. In her book *Exoplanet Atmospheres*, Sara Seager lists four main features of an 'ideal biosignature gas', namely that it doesn't 'exist naturally' on a planet, isn't created by either geophysical or photochemical processes, and has a strong spectral signature. She notes that 'Earth's robust biosignature gas $O_2$ satisfies all four' of these criteria.

The study of exoplanet atmospheres is at an early stage. We have some information, but not a lot. And as yet we have no biosignatures, though that's hardly surprising since most of the planets that have been examined in this respect have been hot Jupiters. What little information we do have will be dealt with in a later section of this chapter. But before we get there let's have a look at the variety of atmospheres that are found in the solar system. Although we only have one life-bearing planetary atmosphere to examine (the usual 'sample size of one' problem), we do have six examples of atmospheres on planets that, as far as we know, do not host life. That's a high enough number to give us a good idea of the range of variation from planet to planet, which turns out to be rather pronounced. This is important, because it means that there is no 'standard' atmosphere for a planet that lacks life – rather there is a wide spectrum of possibilities. Whether this will also turn out to be true of life-bearing planets we don't yet know.

Notice that the number of local non-life-bearing planets given above was six rather than seven. This is because Mercury does not really have an atmosphere; it has such a diffuse cloud of atoms and molecules surrounding it that in Earth terms we would consider it to be a vacuum – or at best as an equivalent to our own rarefied outermost exosphere. However, in addition to the six planetary atmospheres, we also have information on one thick local lunar atmosphere: that of Saturn's moon Titan. We'll take this on board as part of our grand atmospheric tour of the solar system, which is just about to start.

# Earth's Atmosphere

The composition of the Earth's atmosphere has been known in detail for many decades. Indeed, its two main components – nitrogen and oxygen – have been known since the eighteenth century. And the 'noble gases' (such as argon and neon) were discovered before the end of the nineteenth century. These make up a very small fraction of Earth's atmosphere, as do carbon dioxide and water.

It's important at the outset to note that the composition of the atmosphere is a variable thing, both in time and in space. Spatially, it changes with altitude and, to a lesser extent, with latitude. Over time, it exhibits both smallish short-term (e.g. seasonal) changes and much greater long-term (millennia and beyond) changes, notably the *very* long-term increase in oxygen concentration that we noted previously. However, in order to compare Earth's atmosphere with those of other planets in the solar system, it helps to take as a standard measure the atmosphere as it is now rather than in the past, and as it is close to the Earth (the troposphere; up to about 10 kilometres depending on latitude and temperature) rather than higher up. This 'sphere' contains about three-quarters of the total mass of the atmosphere of our planet.

A recurring theme as we scan across the various atmospheres of the solar system is that in each case two gases make up at least 98% of all the constituent molecules. On Earth, these are nitrogen ($N_2$, 78%) and oxygen ($O_2$, 21%), which together make up 99%. None of the other atmospheres in our system come anywhere near to this composition; none has a significant amount of oxygen, and only one (Titan) has a significant amount of nitrogen. So, if we find an exoplanet with an appreciable level of oxygen in its atmosphere, this will be very suggestive of the presence of life, although it would not be conclusive proof.

The composition of Earth's atmosphere is usually given in terms of 'dry air'. This is because the fraction of the atmosphere that is made up of water vapour is very variable – from less than

0.1% to as much as 5%. Water vapour is also found in other local planetary atmospheres, so water is by no means unique to Earth as a vapour, even though it is unique to Earth – within the solar system – as a present-day surface liquid. We might well expect to find water vapour in the atmospheres of exoplanets that are within the habitable zone, as these are likely to have oceans on their surfaces, and oceans are almost certain to be accompanied by atmospheric water vapour.

Our list of components of the Earth's atmosphere so far does not include ozone or methane (which are important in studying exoplanet atmospheres, as we'll see in the section of this chapter entitled *The Magic of Spectroscopy*). There are two reasons for these omissions. First, the level of either of these gases in the atmosphere as a whole is vanishingly small; they are thus described as trace gases. Second, the peak of ozone concentration occurs above the troposphere, at a height of between 20 and 30 kilometres, in the lower part of the stratosphere. Although its existence is very important for life, in blocking much ultraviolet radiation (as we saw in Chapter 4), the fraction of ozone at this altitude in the atmosphere is lower than the name 'ozone layer' implies. In fact, only about 10 parts per million (0.001%) of the ozone layer is composed of ozone. The reason why it is very efficient at absorbing UV radiation despite there being so little ozone present at any one moment in time is twofold: first, each ozone molecule ($O_3$) is a very effective absorber of UV; and second, there is rapid turnover of ozone, so that although it is converted into $O_2$ and O on absorption of UV light, this is rapidly followed by recombination of $O_2$ and O to give $O_3$ again. This dynamic equilibrium is the essence of what's called the ozone–oxygen cycle, which has been estimated to produce about 400 million tonnes of ozone per day.

Several features of the composition of Earth's atmosphere are important to life. Clearly, the presence of a large fraction of oxygen is crucial – at least to the present-day flora and fauna. The presence of water vapour is also important – life does not thrive in ultra-dry environments. The ozone layer provides a useful UV shield. Carbon dioxide is necessary for photosynthesis.

And the presence – in small amounts – of greenhouse gases helps to keep Earth's average surface temperature at its current level. The main greenhouse gases in Earth's atmosphere are water vapour, carbon dioxide, methane, nitrous oxide, and ozone. Although greenhouse gases are often regarded in the popular press as 'bad', because they cause global warming, they are also 'good' because we need them (but not too much of them) to keep us from freezing. The Earth's average surface temperature is approximately plus 15 degrees Celsius (+59 °F), whereas if there were no greenhouse gases in the atmosphere it would be close to minus 20 degrees Celsius (–4 °F).

Although we don't so often think about it, the absence of certain gases is just as important to life as the presence of others. In some cases, this issue is life-form-specific. A good example of this was the toxicity of oxygen to many anaerobic organisms, which caused major extinctions of these after the Great Oxygenation Event some two billion years ago. In other cases, the issue is across the board. For example, if terrestrial life-forms were subjected to high levels of hydrogen cyanide, this would be problematic for almost all of them, regardless of the domain or kingdom to which they belonged. It might seem like an academic exercise to consider the effect on life of having hydrogen cyanide in Earth's atmosphere, because to all intents and purposes there isn't any. But that wasn't always the case. This gas has been detected in interstellar clouds, so it was almost certainly present in our Sun's proto-planetary disc. And to confirm the latter, it has also been detected in comets. It is a trace gas in the atmosphere of Titan. Further afield, it has been detected in an exoplanetary atmosphere (that of 55 Cancri e).

Exactly which and how many presences and absences of particular gases in the atmosphere are required for life to exist is an open question. This is because the chemistry of extraterrestrial life is itself unknown. But sticking to the policy of surmising that biology elsewhere will usually be broadly similar to that on Earth – simply because chemistry, like physics, applies across the universe – we should proceed on the basis that the atmospheric requirements for life on other planets are not too different from those that apply here. Then again, we need to remember that the Earth's atmosphere has changed a lot in the long term, and yet life has survived. Clearly, there has never been a time during the last four billion years in which the Earth's atmosphere was lethal for all forms of life. If it had been lethal, even for a single generation, we wouldn't be here to discuss these matters.

## Venus and Mars

While the composition of Earth's atmosphere was worked out a long time ago, the atmospheres of Venus and Mars have only been known in detail since the 1960s, when the first flybys of these planets took place (*Mariner 2* – Venus 1962, *Mariner 4* – Mars 1965, and many follow-ups). Now we have lists of their component gases, together with information on relative abundances. In one way, their atmospheres are similar to that of Earth, while in another – more important – way, they are very different from ours. The similarity lies in having two principal gases that together make up about 98–99%, and a smallish number of other gases that are present at measurable levels. The difference lies in the identity of the gases.

Both Venus and Mars have an atmosphere that is about 96% carbon dioxide, 2–4% nitrogen. But this statement makes their atmospheres sound more similar to each other than they actually are. The difference is partly in terms of other gases. For example, Mars's third (or equal second) atmospheric component is the inert gas argon (1–2%), while Venus's is sulphur dioxide (less than 0.1%). It's the sulphur dioxide in the atmosphere of Venus that is responsible for the occurrence of sulphuric acid rain there. But the difference between Mars and Venus is more in the density of the atmosphere than in its composition.

So far, we've simply classified planets as having an atmosphere or not, with Mercury being the sole local example of the latter. However, instead of thinking of the binary variable presence/absence, we can think instead of a continuous variable – atmospheric pressure. To do this, naturally, we need some form of measurement. In this respect we're spoilt for choice, as there are many scales on which pressure can be measured. For astronomical purposes the most appropriate of these is the scale based on the unit called the 'standard atmosphere' (or simply 'atmosphere', abbreviated to atm). One atm is the atmospheric pressure at sea level on Earth.

So, by definition, the atmospheric pressure in Manhattan is 1 atm. The atmospheric pressure at the top of Mount Everest is only about a third of this figure – about 0.33 atm. However, such terrestrial variation pales into insignificance when we look at the average pressures on Venus and Mars. To an order-of-magnitude approximation, the atmospheric pressure on Venus is about 100 atm; that of Mars is less than 0.01 atm, making Everest look like it has an amazingly dense atmosphere, despite the fact that humans have difficulty climbing the highest part of the mountain without a supply of oxygen. If we compare Venus and Mars with each other rather than with Earth, the differential in their atmospheric pressures is about 10,000 times. (Even these differences pale into insignificance when we look at the situation on the Moon or on Mercury; on these bodies, the pressure is way less than a trillionth of 1 atm, which is why they are usually said to lack an atmosphere.)

Would the atmospheres of Venus or Mars give any signs of life to a distant observer? The short answer to this question is 'no'. However, Mars has both oxygen and methane as minor atmospheric components, and that combination might be suggestive of life processes if the amounts of those gases were higher than they actually are. Both planets have some water vapour in their atmosphere, but again at a very low level. So an alien intelligence looking at the atmospheric signatures of these planets from afar might well conclude – if they are carbon-based and oxygen-breathing aliens – that neither planet is suitable for life.

## The Strange Case of Titan

Just as Mercury is the odd planet out in terms of not having an atmosphere, Saturn's moon Titan is the odd moon out for having one. Elsewhere in the solar system, moons are much like our own Moon in lacking a dense coating of gases. Titan not only has such a coating, but it has a unique one that resembles neither those of the local rocky planets nor those of its host planet Saturn or any of the other giants (Jupiter, Uranus, Neptune).

Saturn and Jupiter have rather similar atmospheres. Both have two main gases – hydrogen and helium – making up more than 99%. And both have traces of other gases, such as methane and ammonia. Titan's atmosphere again exemplifies the 'two principal components' model, but this time they are nitrogen (98%) and methane (1.5%). Strangely, then, within the solar system, Titan has the most similar atmospheric composition to Earth in the sense that nitrogen is the principal gas in both cases. The gases present at lower concentrations in Titan's atmosphere are also of interest. They include a range of organic molecules that are more complex than methane, in that they have two or more carbon atoms. These include ethane, ethene, ethyne, propane, propyne, and benzene. They also include a number of molecules with a cyanide (CN) component: hydrogen cyanide, cyanogen, and cyanoethyne. Further, Titan is covered by a haze that is made up of a complex mix of organic molecules called tholins.

So far, this description of Titan's atmospheric composition has only been given in the form of lists of gases. Such lists provide a good starting point, in that they reveal both similarities and differences between the atmosphere of Titan and the atmosphere that we're most familiar with – that of Earth. However, lists are ultimately limited in the kind of information they can convey. In particular, they do not in themselves tell us about the dynamic nature of the atmosphere, which is driven by, among other things, photochemical reactions, including photodissociation, in which sunlight – often the UV component in particular – causes the breakdown of atmospheric gases such as methane and ammonia, which is the first step to the production of others. Breakdown of earlier ammonia is thought to be the source of Titan's nitrogen-rich atmosphere; and breakdown of some of the molecular nitrogen ($N_2$) into nitrogen atoms (N) can result in the eventual production of nitriles (cyanide and its cousins).

Photo-dissociation of methane ($CH_4$) is also important. It can be followed by the dissociated parts (for example the highly reactive CH radical, as well as $CH_2$ and $H_2$) getting together in various ways. As well as forming 2-carbon molecules like ethane, multi-carbon long-chain hydrocarbon molecules are also formed. We don't know exactly *how* long, in the sense of how many carbons, though some may run into the hundreds. These contribute to the tholin haze; and some may be present also in the surface lakes.

The complexity of Titan's atmosphere has stimulated many modelling studies, which extend our understanding one stage further. If our first stage was a list of constituent gases and the second stage a broad understanding of the photochemical reactions that take place in the atmosphere, this third stage involves quantification of reaction rates. One outcome of a knowledge of the nature of the photochemical reactions and a prediction of their likely rates is that we would expect the methane in Titan's atmosphere to have dropped to a negligible concentration within the first 0.1 billion years of its history. In contrast, what we actually find now, 4.5 billion years into that history, is that methane is the second-commonest gas, after nitrogen. Since we know that methane is being rapidly destroyed, the only way it can remain present in non-negligible concentrations is if there's an input of methane into the atmosphere to balance the destruction. Although Titan's lakes may provide some methane to the atmosphere, the main source of renewed atmospheric methane is thought to be cryovolcanic activity, transporting methane from deep down within the interior.

Let's return to those long-chain tholin hydrocarbons before leaving Titan. These are among the most complex extraterrestrial organic compounds found to date. It's worth considering how they relate to the long-chain organic macromolecules of life on Earth, in terms of both size and specificity. If some of Titan's hydrocarbons do indeed have 100+ carbon atoms, as currently thought, these are perhaps extending into the size range of proteins. Most proteins consist of a tangled chain of between 100 and 1000 amino acids – though a few are outside this range (both shorter and longer). To make a comparison with the length of tholins, we need to have a rough idea of the number of carbons per amino acid. A simple amino acid like glycine has two carbon atoms. Others have more than 10 carbons (e.g. tryptophan, 11). If we take 5 as a rough average, then a protein with 100 amino acids has about 500 carbons. So in terms of the number of carbons, some Titanic tholin constituents may rival small proteins.

However, when it comes to specificity, things are very different. As we saw in Chapter 13, each protein has a different sequence of amino acids from every other protein. These differences stem from the specificity of the base sequences of the genes that make the proteins. In contrast, some (saturated) long-chain hydrocarbons are simply a recurring sequence of $CH_2$ units, while others (unsaturated) have some variation along their length, in that some links have carbon–carbon double or treble bonds. While the latter (unsaturated) arrangement could provide a degree of sequence specificity, this would be as nothing compared to what is possible with DNA or protein. It would probably not be sufficient to provide a basis for life.

## The Magic of Spectroscopy

On Earth we can directly observe, measure, and test the atmosphere on a day-to-day basis. For some other bodies in the solar system (e.g. Mars, Titan), we can obtain data on their atmospheres from spacecraft, including landers that can sample the atmosphere surrounding them directly. For exoplanets, direct sampling is a complete impossibility, both now and in the foreseeable future. So, for these, we have to make do with information obtained from the techniques of spectroscopy. This means looking at the light arriving here from the system concerned and somehow deducing from that the nature of the atmospheres of its constituent planets. Although this sounds like magic, especially given the distances of many light years that are involved, it's very real science. And we need to know how it works – hence this section.

The basic idea is that the atoms and/or molecules in the atmosphere of an exoplanet 'sign' the light that leaves the planetary system concerned; and we can detect their signatures with certain powerful telescopes, both some that exist already and, importantly, some that are now in the planning and design phase. This atmospheric signing typically takes the form of an absorption spectrum, as described below.

For many planetary systems, the light from the host star reaches us without being influenced by the planets concerned, because they orbit in a plane whose angle is such that they never come between the host star and the Earth (i.e. they never 'transit', from our perspective). In other systems, where the orbital plane is such that the planets do indeed transit, they are often off to the side, from our point of view, and so again we don't see their signatures. But in this latter type of system, during the planet's transit across the front of the star its atmospheric gases absorb some of the starlight, thus reducing the amount that reaches us. Although the overall effect is small, it is very specific to certain wavelengths. And which wavelengths are affected depends on which gases are present. So, for example, the signatures of methane and carbon dioxide molecules are different from each other (Figure 15), and different from the signatures of any other gases. It is just such differences that may provide our first ever indications of extraterrestrial life.

One reason we see signatures of this sort is that atoms and molecules have a series of electronic energy levels, or states, the lowest of which is called the ground state. For any particular atom or molecule, an electron can jump from a lower to a higher state by absorbing a photon of light. Photons with different wavelengths have different energies; and the transition from one electronic energy state to a higher one in the case of a particular atom will correspond to a particular photon. Although the atoms that absorb photons and become excited typically re-radiate the photon a short time later, while returning to a lower state, this photon emission occurs in all

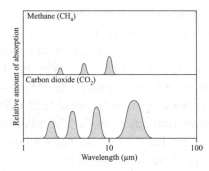

**Figure 15** The pattern of light absorption by the atmospheric gases methane and carbon dioxide, shown in simplified form. The peaks of absorption for one gas are different from those of the other, which enables us to distinguish them in spectroscopic data. The vertical axis is not labelled with units as it gives just relative amounts of absorption. The horizontal axis gives wavelengths in terms of micrometres, and is on a log scale. (One micrometre, alias one micron, is a millionth of a metre.) The complete range of wavelengths given is within the infrared part of the spectrum.

directions. Thus over a period of time, the absorption from the light that's headed in our direction is mostly emitted in other directions, and so we still see the initial absorption as it was, only very slightly reduced by the tiny fraction of the re-emission that happens to come our way.

Spectroscopic analysis of planetary atmospheres can be carried out using various sections of the electromagnetic spectrum – usually either visible or infrared. If we're looking at the visible light from a star, we generally see signatures caused by excitations that are due to electrons jumping to higher energy levels, as discussed above. However, if we're looking at the infrared component of starlight, we're generally seeing signatures caused by the bonds between atoms within a molecule vibrating between higher and lower energy states. The reason for this connection between the type of light (visible, infrared) and the type of energy state transition (electronic, vibrational) is that the differences between vibrational energy states are typically much lower than those between electronic ones, and they thus correspond to photons of lower energy (longer wavelength).

There are limitations to the apparent magic of spectroscopy. Not all molecules have vibrational energy signatures. Whether they do or don't depends on their structures. Specifically, asymmetric molecules have vibrational signatures, but symmetric ones don't, because the signature is associated with a change in the way an asymmetric electrical charge is distributed over the molecule. Thus diatomic molecules in which the two atoms are different, such as carbon monoxide (CO), leave signatures on infrared light, whereas those in which the two atoms are the same, such as nitrogen ($N_2$), typically do not.

We can see a potential problem here. Oxygen is of special interest to us because of its connection with life. But oxygen, like nitrogen, is a homodimer. So, how do we detect it when using infrared spectroscopy? One way we can do this is to infer its existence from the detection of ozone. Since ozone ($O_3$) is a triatomic molecule, it is asymmetric and hence leaves an infrared signature. Photochemical reactions in the atmosphere produce ozone from oxygen, as we saw earlier. If there is no oxygen, there will be no ozone. Hence the signature of ozone can be interpreted as a proxy signature for oxygen. However, that said, oxygen signs light in such a way that its absorption bands span the visible/infrared divide – a divide that is an artificial human construct anyhow in a continuous sequence of wavelengths. There are oxygen bands in the shorter-wavelength part of the infrared, the part that is nearest to visible, and so is called the near-infrared. Of particular interest as a possible biosignature is the oxygen A-band, at about 760 nanometres. Ignas Snellen and colleagues, writing in 2013, suggested that this and another oxygen band may be detectable with the in-construction Extremely Large Telescope (ELT), scheduled for 'first light' in 2025. And when advanced direct-imaging space telescopes (such as the planned HabEx and LUVOIR; see next section) are operating, they will provide another means of detecting oxygen bands in exoplanet atmospheres.

As well as being interested in the occurrence of particular gases in the atmospheres of exoplanets, we are also interested in certain combinations of gases, especially combinations which suggest that the composition of the atmosphere cannot be explained solely in terms of chemical equilibrium. One combination that's of interest in this respect is oxygen/ozone and methane. These react chemically to produce carbon dioxide and water. If analysis of an atmosphere shows that there are significant concentrations of both oxygen/ozone and methane, this suggests that there is a constant renewal of these gases, which is compensating for their removal. One plausible hypothesis to explain this renewal is the existence of a biosphere on the planet concerned, involving, among other things, oxygenic photosynthesis.

## Exoplanet Atmospheres

Analysis of exoplanet atmospheres is a new branch of science. It didn't exist before the current century. The first investigation of a planetary atmosphere beyond the solar system was reported by David Charbonneau and his colleagues in 2002. This was a study of the hot Jupiter HD 209458 b, orbiting very close to its Sun-like host star, which is about 160 light years from Earth. The participants in this study detected atomic sodium in the atmosphere. While atmospheric sodium is not particularly relevant to the search for life (even though sodium is found in all terrestrial life-forms), the demonstration that analysing the atmospheres of exoplanets was possible at all was a significant milestone in exoplanetology.

Since 2002, many other studies have been undertaken of exoplanet atmospheres. One of these was conducted by Jeremy Richardson and colleagues in 2007 on the same hot Jupiter on which sodium had already been discovered. Other studies have typically also been conducted on hot Jupiters, which are unlikely to be hosts for life – at least as we know it. Gases that have been discovered in the atmospheres of one or more exoplanets include: carbon monoxide, carbon dioxide, methane, and water vapour. In 2014, the first result for a smaller planet (HAT-P-11 b, which is approximately the size of Neptune) was published by Jonathan Fraine and his colleagues – and water vapour was seen to be a component of its atmosphere.

The first results for a 'super-Earth' exoplanet, 55 Cancri e, which has a mass approaching 10 times that of Earth, were published by Angelos Tsiaras and his colleagues in 2016. These showed the existence of atmospheric hydrogen and helium. 55 Cancri e is the closest-in planet of a system that contains at least five of them. It orbits at a distance of less than 0.02 AU from its K-class sun – more than 20 times closer than Mercury is to our own Sun. And its orbital period is a mere 18 hours. This is no more likely to host life than is a hot Jupiter or hot Neptune. In 2019, Tsiaras and colleagues found water vapour in the atmosphere of another super-Earth, K2-18 b. Unlike 55 Cancri e, this planet appears to be in the habitable zone of its host star, which is a red dwarf.

At the time of writing, the first detailed spectroscopic studies of the atmospheres of Earth-sized exoplanets in habitable zones have yet to be undertaken, though some preliminary observations of the atmospheres of four of the TRAPPIST-1 planets were published in 2018 by Julien de Wit and colleagues. Eventually, some such detailed studies may provide evidence of oxygen/ ozone, perhaps in combination with detectable levels of methane. Such a result, if and when it comes in, will be hugely important. Just as the discovery in 2014 of Kepler-186 f – the first quasi-Earth – was a major milestone, the first quasi-Earth-like atmosphere will be a major milestone too. When we reach it, we just need to await the third major milestone – the detection of the first biosphere and its constituent creatures.

What type of telescope will be used to provide our first evidence of atmospheric gases – or combinations of gases – that will suggest the presence of life? The Kepler Space Telescope gave us lots of exoplanet finds, and its successor TESS will continue to do this for several years; but these were designed primarily to detect the planets themselves, not to detect the gases that make up their atmospheres. The study of atmospheric compositions on Earth-like exoplanets requires a new type of space telescope, one that operates via direct imaging – that is, seeing the planet itself. This is very different from indirect imaging – inferring the existence of a planet by observing its effects on its host star's light or pattern of movement.

Future systematic searches for biosignature gases like oxygen and ozone in the atmospheres of smaller Earth-like planets will be carried out by a new generation of direct-imaging space telescopes that are currently being planned to specialize in this activity. The two most exciting planning-and-design projects at present are HabEx and LUVOIR. The former is short for Habitable Exoplanets Imaging Mission; the latter is an acronym for Large Ultra-Violet Optical Infra-Red Surveyor. The huge advantage of direct imaging is that it allows atmospheric analysis of non-transiting planets. These constitute the vast majority of planets, and they are invisible to space telescopes like Kepler and TESS, which search using the transit method. The probability of an Earth-like planet orbiting a sun-like star exhibiting transits for an Earth-based (or Earth-orbiting) viewer is about 0.5%. Thus for every Earth-like planet that Kepler or TESS sees, a direct-imaging telescope has the potential to see about 200.

There are two major problems in attempting the direct imaging of exoplanets. The first is that the light from a planet is swamped by the light from its much brighter host star. At first sight, this problem would seem to be insurmountable. Typically, a star will outshine one of its orbiting planets in terms of visible light by a huge multiple – billions of times. For example, the Sun outshines the Earth by a factor of 10 billion (or $10^{10}$). If we look at an exo-Earth orbiting an exo-Sun say 50 light years away, how can we possibly hope to see the planet at all?

A solution to this problem is to search in the infrared waveband, where the contrast ratio between sun and planet is less pronounced than it is for visible light. For example, the Sun 'only' outshines the Earth by a factor of 10 million ($10^7$) in the longer-wavelength part of the infrared region (known as the thermal infrared because we feel it as heat). While this is still a pretty adverse ratio, it is a thousand times better than the one that applies in the case of visible light.

Unfortunately, while this long-wavelength approach eases the 'swamping' problem, it exacerbates a second major problem – that of distinguishing, or spatially resolving, the radiation emitted by a star from that emitted/reflected by an orbiting planet.

The longer the wavelength of the light we observe, the harder this resolution becomes, and hence the bigger the primary mirror (or lens) of the telescope that is being used needs to be in order to achieve it.

Here's an example to illustrate the problem – for a fuller version, see James Kasting's 2010 book *How to Find a Habitable Planet*. To resolve the radiation from a quasi-Sun–Earth system that's about 30 light years away from ours, a telescope using light at the boundary between visible and infrared would need to be about 8 metres in diameter, while a telescope using thermal infrared light would need to have a diameter 10 times greater. The former is in the doable range: each of the four mirrors of the Very Large Telescope (VLT) in Chile is 8.2 metres in diameter, and the planned combined mirror diameter for the James Webb Space Telescope is 6.5 metres. In contrast, the latter – a mirror of about 80 metres in diameter – is not yet in the range of the doable. However, this does not mean that we should give up on the idea of operating with thermal infrared wavelengths. If we use a pair (or array) of space telescopes, the distance between the telescopes can be the critical factor rather than the diameter of the mirror in each – and the former is of course the greater of the two. In such a setup, we look at the way in which the infrared radiation received at the two mirrors combines – this is called interferometry.

A different approach is as follows. The problem of swamping of the planet's radiation by that of the star in the visible section of the electromagnetic spectrum can be alleviated by physically blocking the light from the star. There are various ways to do this, including a built-in shield, as used by the SOHO spacecraft (Solar and Heliospheric Observatory, launched in 1995 and still operating today), called a coronagraph. By blocking light from most of the disc of a star, it is possible to look at the light from its outer corona or the light from an orbiting planet. The same thing could be achieved by flying a separate spacecraft (a large-diameter 'starshade') in between the space telescope and the system it is observing – plans for such starshades are under development.

The choice of waveband in studies of exoplanet atmospheres is clearly crucial. As currently envisaged, both HabEx and LUVOIR will be able to observe visible and near-infrared light (so both would be able to see the important oxygen A-band). Below and above this range of wavelengths the two proposed telescopes differ somewhat. However, the current designs will probably change; indeed the existing LUVOIR proposal already features alternative designs. Whether either of these two missions ever takes off will probably be decided in the next year or so, by the upcoming decadal surveys to be produced for NASA; even if they are successful in being funded, their spectroscopic analyses of exoplanet atmospheres probably won't begin until the 2030s.

In the meantime, one thing we should focus on is the possibility of obtaining false positives and/or false negatives in our search for life. A significant level of oxygen would be exciting but not conclusive if some abiotic process could be imagined that would be able to sustain such a level, so it could be a false positive. Lack of a 'red edge' (a spectroscopic signature of chlorophyll in Earth's atmosphere) could be a false negative in the case of an exoplanet – because other photosynthetic pigments might be commoner there than they are here. The more we have considered the possible ways of interpreting various spectral signatures by the time we start obtaining the relevant data, the better equipped we will be to undertake this hugely exciting endeavour.

# 16 RADIO AND LIFE

## Homage to James Clerk Maxwell

Some scientists argue that the third most important physicist of all time, after Newton and Einstein, was the Scotsman James Clerk Maxwell, who was born in Edinburgh in 1831. Maxwell lived a short but unusually productive life. He died in 1879 at the age of 48 from the same disease – stomach cancer – that his mother had died from, also at the age of 48. While lifestyle factors are important in relation to the incidence of cancer, so too are genetic ones.

In 1865, while he was a professor at King's College London, Maxwell published a paper entitled 'A dynamical theory of the electromagnetic field' in the journal *Philosophical Transactions of the Royal Society of London*. This brought together electricity, magnetism, and light. It was followed by a two-volume book, *A Treatise on Electricity and Magnetism*, published in 1873 by the Clarendon Press. At this time, Maxwell was professor of physics at Cambridge University; he was also spending a good deal of time at his family home at Glenlair, near the village of Corsock, in the (current) southwest Scottish region of Dumfries and Galloway.

Among other things, Maxwell showed that electricity and magnetism were complementary aspects of something that encompassed them both – electromagnetic fields. He showed that electromagnetic waves travelled through space at the speed of light, and thus determined that visible light was a form of electromagnetic radiation. This scientific leap led in stages to today's understanding that the things we call gamma rays, x-rays, ultraviolet, visible light, infrared, microwaves, and radio waves are all forms of electromagnetic radiation, but with different wavelengths.

The symbol for wavelength is λ, the Greek letter lambda. The light that we can see has lambda values of between about 400 (violet) and 700 (red) nanometres (nm; one nanometre is a billionth of a metre). As the names suggest, ultraviolet and infrared radiation flank this region, with ultraviolet rays having wavelengths between 100 and 400 nm, and infrared being from 700 nm up towards 1 millimetre, at which point we call the rays microwaves. The range for microwaves spans from about 1 millimetre to 1 metre, after which point we are into the radio section of the electromagnetic spectrum.

The radio section is extremely broad, so it is split into various subsections, many of which are familiar to us from technical descriptions of radio channels. These are typically given in terms of the frequencies of the radio waves used, for example VHF (very high frequency). The frequency of a wave is the number of cycles per second. One cycle per second is a frequency of 1 hertz (by definition), and one million cycles per second is 1 megahertz (MHz). VHF radio waves have frequencies between 30 and 300 MHz.

But how does this relate to wavelengths? If the radio spectrum has wavelengths from about 1 metre upward, how do sections of this spectrum in wavelength terms relate to sections like VHF, which are normally defined in terms of frequency? In fact, the answer is straightforward, because wavelength and frequency are inversely related to each other. The speed at which a wave is travelling is the length of the wave (in units of distance, like metres) times the number of them that go past a fixed point per second. Or, speed = wavelength × frequency. We know that electromagnetic waves travel at light-speed; so, if we know the frequency of a wave we can work out its length. Doing this for VHF, we find that its wavelength range is from about 1 to 10 metres. For VLF (very low frequency), the wavelength range is from about 10 to 100 kilometres (Figure 16).

Note that there is much use of the word 'about' in the above account. We should recognize that the electromagnetic

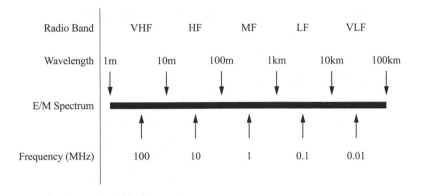

| Radio Band | VHF | HF | MF | LF | VLF |
|---|---|---|---|---|---|
| Wavelength | 1m | 10m | 100m | 1km | 10km | 100km |

E/M Spectrum

| Frequency (MHz) | 100 | 10 | 1 | 0.1 | 0.01 |

**Figure 16** The relationship between wavelength, frequency, and radio bands (VHF etc.). Note that wavelength increases from left to right, while frequency decreases. Frequency is given in units of megahertz (MHz). One MHz is a million hertz (Hz), and 1 Hz is defined as 1 cycle per second. Continuing the electromagnetic spectrum to the left of the section shown takes us to microwaves, infrared, and then to visible light; while continuing to the right leads to ever-longer radio waves.

spectrum is, as its name suggests, continuous. There are no abrupt thresholds at particular points along it that mark the distinction between, for example, microwaves and radio waves. Thus our series of categories described by words like ultraviolet, visible, infrared, microwave, and radio are categories of convenience. Some of them do have fairly definite limits from our perspective as biological entities receiving them. But even in these cases, like the start and end of the visible-light category, the limits concerned are literally in the eye of the beholder. For a bee, the limits of the category 'visible light' are different from those for a human.

As an aside, the hertz unit of the frequency of a wave is named after the German physicist Heinrich Hertz, who proved Maxwell's electromagnetic theory through a series of experiments conducted in the 1880s. These were designed to detect electromagnetic radiation with a frequency of a few metres – radio waves as we now call them. His success demonstrated that Maxwell's theory was correct. Hertz had an even shorter lifespan than Maxwell – he died at the age of 36, due to a condition that involved inflammation of blood vessels within the organs, the cause of which was unclear.

## Radio Waves and Biology

Electromagnetic waves are of the utmost importance to living organisms. But which wavelengths are important, and which are not? The very shortest (gamma and x-rays) are highly damaging and for the most part it's their absence that's important rather than their presence. An obvious exception is human use of x-rays for medical purposes – but note how important it is that radiographers are shielded from the rays. Ultraviolet rays are important too; and they cross the divide between harmful and helpful. As we've seen, the ozone layer blocks much of the shorter-wavelength ultraviolet from reaching the Earth's surface. But for the most part the longer-wavelength ultraviolet gets through to the surface (clouds permitting), and it helps humans to make vitamin D. Also, there are reflection patterns of ultraviolet light from the petals of flowers that are visible to some insects – but not to mammals, including humans.

The importance of visible light to organisms is hard to overstate. It powers photosynthesis and thereby most of the biosphere. Plants absorb it; most animals see it. Huge amounts of evolution have happened in direct relation to it, including the production of both cryptic and warning pigmentation patterns. Some life-forms even produce light. This is seen in the bioluminescence of a wide variety of organisms, including vertebrates, invertebrates, fungi, and bacteria. Vertebrate examples include the appropriately named lantern fish; invertebrate examples include fireflies and glow-worms, both of which are actually beetles ('flies' being adult beetles and 'worms' being larval ones). There appear to be no bioluminescent plants, except for some in which the luminescence is caused by symbiotic bacteria (this also occurs in some animals), and some in which it is the result of bio-engineering.

Moving into the longer wavelengths of the electromagnetic spectrum than those of visible light takes us first to infrared.

Naturally, this is ubiquitous among life-forms on Earth, just as it is among inert objects, since all bodies (of whatever construction) emit infrared radiation as long as they are at a temperature that is higher than absolute zero. However, some animals can actually 'see' infrared radiation. A good example of this is found in snakes, where it has evolved multiple times, including in the boa, viper, and python families. These snakes can target their prey by detecting infrared wavelengths between about 700 and 3000 nanometres.

Whether 'see' is the right verb is a moot point. Certainly the snakes concerned can perceive infrared, but how do they do so? Although the details vary a little from family to family, the basic method does not. These snakes have organs called loreal pits that are close to their eyes. Even in the most developed of these – in the pit vipers – there is no retina. However, nerve signals from the pits go to the same area of the mid-brain as do signals from the eyes. It is thought that in this area a sort of combined visual and infrared map is built. This allows the snake to locate its prey using both types of information. The infrared 'vision' may also be useful to the snake in seeing potential predators.

Vampire bats can also sense infrared – they are the only mammals known to be able to do so. There are three species of these, which make up a small part of the family Phyllostomidae, or New World leaf-nosed bats – the family has about 200 species overall. All three vampire species feed on mammalian and/or avian blood, as their name suggests. Their ability to detect infrared radiation almost certainly enables them to find the parts of the body of their prey that have sizeable blood vessels running nearest to the skin. The details of their detection system are less well known than in the snakes. However, it's thought that the organs responsible are in the leaf-shaped nose, and that the signals are conveyed to the same part of the brain as in their snake counterparts.

Although we humans cannot see infrared radiation, we can feel it as heat, and we use it for many purposes. In addition to its use in spectroscopy, it's used to steer heat-seeking missiles; and it's even used in some forms of communication, including the 'talking signs' system that can be used by people with visual impairment.

But what about the longest-wavelength section of the electro-magnetic spectrum, namely radio? This is clearly very useful to modern humans. But is it used or produced by any other species of organisms here on Earth? The answer would appear to be 'no'. But this in turn leads to another question: why not? In other words, since evolution has been able to make some animals that see ultraviolet light, and others that see infrared light, why has it never, in any of those countless millions of lineages, over many millions of years, made animals that see radio waves?

Although we can often make a good stab at answering the sort of question 'why has evolution produced X?', it's usually much harder to answer the opposite sort of question – 'why has evolution not produced X?'. In the case of radio, three possible reasons are as follows. First, there's not much of it about. The Sun is the main local producer of electromagnetic radiation. Its production peaks in the visible part of the spectrum, and falls off rapidly to both sides of this maximum. The same is true of other stars, though the exact position of the peak varies somewhat, in a way that is related to temperature by Wien's law (an amateur version of which is: the hotter the star, the shorter its peak emission wavelength, and so the bluer it is). Second, it's not clear what benefit an organism would get by seeing this very-long-wave-length radiation, since no other organisms emit it, in the same way that they do its shorter-wavelength infrared counterpart. Third, given the long wavelengths involved, animals would need to have unrealistically large eyes to be able to see it.

So, it would seem that the ability to both transmit and receive radio waves is, in terms of the biosphere of Earth, a uniquely human trait. However, it hasn't been a human trait for long. James Clerk Maxwell's theoretical work in the 1860s suggested that there should be such things as radio waves – albeit the term 'radio' wasn't yet used at the time. Hertz's experimental work in the 1880s confirmed that Maxwell was right. But it wasn't until the final years of the nineteenth century that radio communi-cation between one human and another became real. And it wasn't until the first years of the twentieth century that such communication was used over long – for example transatlantic – distances. Let's now investigate the advent of radio communication among humans more closely.

## Human Radio Communication

The key player in the development of early radio was the Italian electrical engineer Guglielmo Marconi, who is often credited as the 'inventor of radio', even though many other people were also involved in the progress that led from Maxwell and Hertz in the late nineteenth century to the start of regular radio transmissions in the 1920s. Marconi transmitted radio signals over short distances (a few kilometres), first in Italy in 1895 and then in Britain in 1897. Long-distance transmission soon followed, though there's some uncertainty over whether Marconi's first claimed transatlantic success (eastward) was real. Be that as it may, the first claimed westward transmission in 1902 is generally accepted. Seven years (1895 to 1902) thus saw the maximum distance of successful radio transmission rise a thousand-fold.

We'll shortly be returning to Frank Drake's approach to estimating the number of civilizations with radio communication in the Milky Way. And we'll need to input into Drake's equation the average duration, in years (or centuries, millennia, or millions of years), of broadcasting civilizations. Of course, this average is unknowable at present and has to be guestimated. And all we have to go on is how long the one broadcasting civilization we know of – our own – will endure. This duration is bounded by the year in which radio transmission started and the year in which it stops. At least we can now pin down the first of these figures to an accuracy of about 50 years (see below), even if the second requires another very risky guestimate.

The earliest that humans could be said to have become a broadcasting species would be around 1900. Alternatively, we might choose to ignore those few early transmissions and instead say that we became a broadcasting species around 1920, when regular radio broadcasts began in several countries:

first the USA, followed closely by Argentina, Australia, and Britain. However, there's a third option. Although early radio broadcasts could have leaked out into space, giving an extraterrestrial intelligence a clue about our existence, the first deliberate beaming of radio messages out into space didn't take place until 1962 – a pioneering Soviet message using Morse code – which was followed, 12 years later, by the Arecibo message (see next chapter). So, there's a 'wobble' of just over half a century in what we might deem to be the start of human radio broadcasting from an alien perspective.

Whether this wobble matters or not depends on how long we continue to exist as a broadcasting species. If for a million years, then it's trivial. However, if for a century (perhaps due to nuclear war or asteroid impact), then it's really rather important. Anyhow, my personal choice is the year 1962, because not only was it the year of the first deliberate radio message sent into space, but also it was the year of launch of the famous Telstar-1 radio communications satellite.

Since we have a start-year but not a stop-year for human broadcasting (and since we don't know the equivalent start and stop figures for other broadcasting civilizations), the best approach (next section) will be to run Drake's equation several times, each time inputting different possible average durations for the broadcasting life of civilizations, and to see what effect varying the duration has on the number of civilizations that are predicted to be broadcasting in the Milky Way right now.

Before leaving the fascinating story of the birth of radio transmission on Earth, let's take a quick look at that early developmental period between 1900 and 1920, when radio transmissions took place, but there was little if any regular broadcasting. A major user of radio in this period was shipping. Ships sent messages to each other, often for safety reasons. It's interesting to look at the *Titanic* disaster from this perspective. This ship was built in Belfast, the city where I was born about 40 years later. It began its sea trials on 2 April 1912. It then sailed to Southampton, on the south coast of England, from where it began its maiden journey to New York on 10 April. After stops at

Cherbourg (France) and Queenstown (now Cobh), Ireland, the ship sailed out into the North Atlantic on 11 April. On 14 April, it received several radio messages from other ships, warning it of icebergs. However, these were interpreted as being of a 'proceed but keep a lookout' nature. *Titanic* hit the fatal iceberg just before midnight, and sank in the early hours of 15 April. Although the radio messages that could have saved it were not acted upon, radio distress messages sent from the ship after the collision were responsible for reducing the death toll by calling for help from other ships in the vicinity, leading to the arrival of the *Carpathia*, which rescued about 700 people (out of more than 2000).

Today, radio messages remain important for shipping. And now they are important for other forms of transport too, notably aeroplanes and spacecraft. Radio is how we found out what happened on the *Cassini* mission to Saturn; it is also how we know what is still happening with the *Curiosity* rover on Mars, and what is the current position of *Voyager 1* as it travels beyond the solar system. In addition, we control all such missions by radio. In short, radio is an integral part of the space programme, and represents our main potential mode of communicating with extraterrestrial intelligence. But is there any such intelligence? Let's see.

## Back to Frank Drake

In Chapter 12, we used part of the Drake equation to estimate the number of planets with microbial life in the Milky Way. We came up with a rough estimate of 2.25 billion; however, we rounded that down to an order-of-magnitude estimate of 1 billion. Rounding downward rather than upward is always a good policy as an antidote to overenthusiasm for finding life. Now let's start from that rounded-down figure and modify it, in the way that Drake suggested, to come up with an estimate for the number of broadcasting civilizations in the galaxy.

To get from microbes to broadcasters, we need to input the probability of life evolving into intelligent life, and the probability of intelligent life becoming broadcasting life. Coming up with these probabilities, even as guestimates, is extremely difficult. One approach would be to say that in either case we have no idea and so an unbiased approach, given a probability scale of 0 to 1, is to choose 0.5. Another approach is to surmise that both probabilities are quite small, and hence to choose a low fraction, say 1% (0.01). Let's do both of these things in the calculation below, so we see how the difference between them influences the outcome.

We also need to input a figure for how long broadcasting civilizations endure as such. If we were starting from scratch here, it would make sense to give this figure in years. However, since in fact we're modifying our earlier figure for the number of planets with at least microbial life, we should instead give the figure as a fraction of the duration of that kind of life. Taking the Earth to date, microbes have been here for about 4 billion years (we'll ignore their likely period of living on into the future for now). If we think that broadcasting human life will remain as such for 4 million years (a typical species lifespan, if there is such a thing), then the fraction is 1 in 1000. Alternatively, we could be pessimists and surmise that broadcasting life will exterminate itself within 4000 years, in which case the relevant fraction is 1 in a million. Let's now use these figures to get some estimates for the number of broadcasting civilizations, and then later examine some problems that for the moment can remain hidden.

Using the most optimistic figures above (probabilities of 0.5 and broadcasting lifespan of 4 million years), our estimate of the number of broadcasting civilizations in the Milky Way is 1 billion × 0.5 × 0.5 × 1/1000, which works out at 250,000. However, using the most pessimistic figures (probabilities of 0.01 and broadcasting lifespan of 4000 years), the calculation is 1 billion × 0.01 × 0.01 × one-millionth, which works out at 0.1 – in other words only one broadcasting civilization for every 10 galaxies like the Milky Way. So, which estimate do we believe?

Not only is the answer to this question not clear, but we need to keep reminding ourselves that we can't operate to more than an order-of-magnitude level of guestimates. So let's say that our degree of uncertainty regarding the number of civilizations in the Milky Way that were broadcasting a century ago ranges from 0 to 1 million. (A range of uncertainty for the present day, rather than a century ago, has its lower bound at one rather than zero, because of our own broadcasting activities.) Now let's remind ourselves of the broader context within which we should view these guestimates. So here is our most recent guestimate in relation to our earlier ones:

Number of planets in the Milky Way: 1 trillion
Number of planets with microbial life: 1 billion
Number of planets with animal life: 10 million
Number of planets with broadcasting life: between 0 and
   1 million

While the estimated number decreases going down the above list, the errors around it increase in proportional terms, due to the inclusion of more parameters with progressively more uncertain values in the equation. We end up with the highly unsatisfactory conclusion that on the basis of current knowledge we cannot distinguish between the possibilities of humans being the only broadcasting civilization in the Milky Way, or just one out of a million or so. Truly, we *may* be alone at this level.

Even if we are alone in the Milky Way, we are almost certainly not alone in the universe as a whole. To get from our galaxy to the universe we need to multiply by at least a trillion (we'll examine this multiple further in Chapter 19, in relation to the cosmological principle). Thus if the number of broadcasting civilizations is one per 10 galaxies, then to an order of magnitude there may be 100 billion broadcasting civilizations in the universe overall – it's just that most of them are too far away for us to entertain reasonable hopes of being able to communicate with them.

## Imagining Radio Eyes

There's an important distinction between radio waves and radio signals. Since the latter will be the main focus of attention in the next chapter, we need to be clear about this distinction, and about what a radio signal has, that a radio wave *per se* does not. To do this, we'll start by thinking about how real eyes receive light, and about how imaginary eyes might receive radio.

Amateur astronomers often use red lights when observing the night-time sky, because they interfere less with our ability to see starlight than do white lights. When a fellow astronomer shines a red flashlight towards me, I receive three main types of information. First, I receive the binary information that there's light coming from a particular location rather than no light. Second, I can roughly assess the amount of light from the brightness that I perceive. Third, I can tell from the fact that I'm only seeing red light that the wavelength is between about 600 and 700 nm.

This is very limited information; and if the person shining the flashlight towards me holds it steady for a long time, there is really no additional information added – except perhaps that the batteries of the flashlight are not too low. However, if my flashlight-shining friend senses a problem, such as seeing a nocturnal hunter heading in our direction armed with a rifle, she can alter something to convey more complex information. For example, she could use the on/off switch to send pulses of light rather than a continuous beam; and perhaps under the circumstances she might choose to send three long pulses, then three short, then another three long ones – the Morse code version of SOS.

Now suppose for a moment that in some parallel universe there are beings who see radio. A radio-flashlight would, like its visible-light equivalent in the actual universe, convey just the same basic information of on/off, strength, and 'colour' (i.e. wavelength). But again the being shining the flashlight could convey more complex information, including whatever equivalent of SOS exists in the parallel universe, by varying something over time; and again the simplest way of doing this is to break the continuous beam up into discrete pulses.

When we change something in an otherwise constant beam of electromagnetic radiation, we are modulating the information that's being conveyed. It is essentially a pattern of modulation that changes a beam into a signal. This applies with both types of light (visible, radio) in both universes (ours and the parallel one). In relation to the radio that we broadcast in our universe, the modulation can take a variety of forms. Commonly encountered are amplitude modulation (AM) and frequency modulation (FM). The precise forms these take aren't important (though there are clues in the names) and there are several other forms of modulation anyhow. The important thing is that modulation is happening. It's often happening in side bands – wavelengths close to, but not the same as, the 'main' one, which is called the carrier wave.

This principle of modulation applies to radio signals regardless of the context. What sort of signals we wish to send differs between, for example, a commercial radio station broadcasting light music and a radio telescope sending information on human DNA to our closet exoplanet, Proxima b. But in both these cases, indeed in all cases of radio *signals*, it is modulation that changes a simple wave-form with a low information content into a more complex one with a higher content. And this principle applies to listening for incoming alien signals, just as it does to sending our own out into the limitless void of space.

# 17 SIXTY YEARS OF SETI

## Pioneers of the Alien

An important paper was published in the journal *Nature* in September 1959 by two physicists, one Italian and one American. It asked the question of whether and how we might communicate with aliens – and in so doing, it ushered in the era of SETI – the search for extraterrestrial intelligence. As in many cases of important papers, it involved collaboration between scientists whose home regions were far apart. As in the case of the Watson–Crick DNA *Nature* paper of 1953, which we looked at earlier, one of the scientists concerned – this time Giuseppe Cocconi – had crossed the Atlantic for the purpose of scientific collaboration.

Giuseppe Cocconi had studied at Milan and Rome, and was working at the University of Catania (founded in 1434) in Sicily when he accepted an invitation to work at Cornell University in Ithaca, New York (founded in 1865). He went to Cornell in 1947 and stayed for more than a decade, returning to Europe eventually to play a major role in the early days of the development of CERN (Conseil Européen pour la Recherche Nucléaire), which was founded in Geneva in 1954. The famous *Nature* paper on interstellar communication was published jointly with Philip Morrison, who was also at Cornell at that time.

Morrison had studied for his PhD at the University of California, Berkeley (founded in 1868), under the supervision of Robert Oppenheimer, an outstanding nuclear physicist who worked on a wide range of subjects from black holes to quantum mechanics, but is perhaps best known for his role in the Manhattan

Project during the Second World War, and is sometimes referred to as 'the father of the atomic bomb'. Morrison too made contributions to the Manhattan Project, but after the war worked on astronomical topics, including cosmic rays and gamma rays. Indeed, he is regarded as one of the founders of gamma-ray astronomy.

The 1959 paper by Cocconi and Morrison begins as follows: 'No theories yet exist which enable a reliable estimate of the probabilities of (1) planet formation; (2) origin of life; and (3) evolution of societies possessing advanced scientific capabilities.' Interestingly, while the same is true for (2) and (3) today, exoplanet studies conducted between the 1990s and now do allow quasi-reliable estimates of the probability of planet formation, as we've seen in earlier chapters. Anyhow, Cocconi and Morrison go on to hypothesize that 'near some star rather like the Sun there are civilizations with scientific interests and with technical possibilities much greater than those now available to us.' They envisage that such civilizations will have set up a signal beamed in our direction (among others) and that they are 'patiently' waiting for us to discover it. They then ask what form this signal might take.

Their proposals are: first, that the signal must be based on electromagnetic radiation; second, that the radio section of the spectrum offers the best prospects; and third, that specifically we should search for signals in the vicinity of 1420 MHz frequency, which corresponds to a wavelength of about 21 centimetres. Such waves are in fact microwaves under the definition that these extend from 1 millimetre to 1 metre; but recall that our labels for sections of the electromagnetic spectrum are somewhat variable in their usage. Microwaves are sometimes thought of as being intermediate between radio waves and infrared. However, they are sometimes alternatively thought of as being a subband within the radio section – indeed the 'micro' comes from a comparison with (other) radio waves, not with the electromagnetic spectrum as a whole. From a gamma ray's perspective, microwaves are very 'macro' in terms of wavelength.

But why look at that particular wavelength/frequency? To answer this question, we need to know a few things about hydrogen. First, it's the most common element in the universe. It makes up about three-quarters of the mass of all 'normal' (i.e. non-dark) matter. It dominates both the interstellar medium and the interiors of stars. Second, it occurs in three main forms: atomic hydrogen (H, one proton plus one electron); ionic/plasma hydrogen ($H^+$, protons and electrons are separate); and molecular hydrogen ($H_2$, two bonded atoms). Other rarer forms exist too, for example the isotope deuterium, or 'heavy hydrogen', in which there is a neutron, as well as a proton, in the nucleus. Third, the relative abundance of the different forms of hydrogen varies from place to place. Inside stars, the predominant form is plasma. In the interstellar medium, the denser 'clouds' have a mixture of H and $H_2$, while the thinner intercloud medium has a mixture of H and $H^+$. Finally, hydrogen atoms have a 'signature' at 1420 MHz frequency (21 cm wavelength), which is referred to as the hydrogen line.

Cocconi and Morrison refer to the hydrogen line as 'a unique, objective standard of frequency, which must be known to every observer in the universe'. So they argue that intelligent aliens would have set up communications at or near this frequency, and it's those that we should be looking for. This is a reasonable argument, but there are some problems. In particular, broadcasts at that frequency may become Doppler-shifted, so we need to look in a band around the 1420 MHz frequency to allow for this, as Cocconi and Morrison suggest. Also, although an artificial signal should be distinguishable from natural emission of radiation of this wavelength by hydrogen atoms, there might be a danger of the signal getting swamped in noise. To overcome this problem, the American scientist Carl Sagan, famous for the TV series *Cosmos* and his linked 1980 book of the same name came up with an ingenious solution. He suggested that when sending out signals of our own we might use not 1420 MHz but instead 1420 either multiplied or divided by a universal constant like $\pi$.

Cocconi and Morrison end their 1959 paper by urging their readers not to consider the paper to belong in the realm of science fiction. And for the doubters of the likely success of a search for radio signals from intelligent aliens, they provide a wonderful concluding sentence: 'The probability of success is difficult to estimate; but if we never search, the chance of success is zero.' This can be seen as an exhortation to the scientific community to engage in a major practical SETI effort – something that happened, and is continuing to happen. Let's now have a look at this endeavour.

## SETI Begins

There had been a few attempts at practical SETI long before the Cocconi–Morrison paper. For example, in 1924 during a period when Mars was unusually close to the Earth, a radio receiver searching at a very long radio wavelength (a few kilometres) was lifted to an altitude of about 3000 metres above the US Navy's observatory near Washington DC, and pointed at the red planet. No Martian messages were received, and it's now clear, given the photos we have of the Martian surface, that we were a little optimistic to have expected any.

However, SETI started in earnest immediately after the Cocconi–Morrison paper. The first search was initiated by Frank Drake in early 1960, under the label Project Ozma. This involved searching for radio signals at the 1420 MHz frequency coming from two nearby Sun-like stars – Tau Ceti (about 12 light years away) and Epsilon Eridani (about 10 light years). The search was conducted from the National Radio Astronomy Observatory, at Green Bank, West Virginia. Neither Ozma, nor its follow-up project Ozma II, found any alien signals. One seemingly promising signal was discovered in April 1960. However, its source turned out to have been a human-piloted aircraft rather than an alien civilization.

Before looking at the ongoing development of SETI from the starting points of the Cocconi–Morrison paper and Drake's Project Ozma, we need to make an important distinction, namely that between incoming and outgoing signals. While the starting point for SETI was a search for the former, the field has expanded to also include the latter. Sometimes outgoing ones are considered under the separate heading of METI (messaging), CETI (communicating with), or 'active' SETI, though I see no need for any of these labels. Both sending signals and attempting to find signals from alien civilizations are part of SETI in a broad sense. So, to avoid an explosion of acronyms, I'll just use SETI for the whole endeavour, but I'll make it clear whether I'm referring to incoming signals, outgoing signals, or both.

With regard to incoming signals, we should really admit that the answer to all our listening so far has been the proverbial deafening silence. The vast majority of the time devoted to this has not produced any evidence of possible alien messages. There have been a few exceptions, some of which are now known to have come from sources other than alien intelligence – such as the aircraft signal in Project Ozma, and the Little Green Men pulsar that we looked at in Chapter 3.

Other interesting incoming signals do not yet have an obvious explanation, but nevertheless seem to be unrepeatable observations. For example, in 1977, the American astronomer Jerry Ehman, working at Ohio State University's Big Ear telescope, received the 'Wow!' signal at a frequency of approximately 1420 MHz from the direction of a star in the constellation of Sagittarius. Ehman circled a string of symbols that appeared in his computer record (which, given the year, consisted of line-printer output): 6EQUJ5. In the margin beside this, he wrote 'Wow!' to indicate the potential interest in this signal from a SETI perspective. However, we need to enquire carefully about what this string of symbols means, because the alphanumeric nature of the computer output could easily be misinterpreted.

In the system Ehman was using, the strength of the radio signal received was represented by the numbers 0 (lowest) to 9 (higher); then letters were used to represent still higher strengths. So A followed on from 9, and was followed by B, C, D, and so on, potentially up to Z. Thus the sequence 6EQUJ5 indicates a spike of signal strength, peaking at U. Repeats of this sudden spike of radio output at a frequency very close to the hydrogen line, coming from this particular direction, have been searched for on many occasions since, but never found. Various hypotheses have been put forward to explain the pulse, but there is still no consensus on its source. Thus it remains a possible candidate for an alien explanation, but a rather weak one.

What have we heard since 1977? The short answer to this question is 'nothing'. Well, nothing that could reasonably be interpreted as an alien message anyhow. We've heard lots of other things, including some that were unknown back in the early days of SETI. For example, FRBs (fast radio bursts) were discovered in 2007, and we've since seen quite a lot of them. Unlike the alien 'open channel' predicted by Cocconi and Morrison in 1959, these bursts are very short (each being a small fraction of a second) and they span a wide range of frequencies. Study of them so far suggests very powerful sources beyond the Milky Way. There are many hypotheses as to their origins, and indeed there may be more than a single sort of source. One possibility is a destructive collision, for example between neutron stars or black holes. However, while most of the FRBs we've seen appear to be one-off bursts of radiation, a few are repeating bursts, which suggests an ongoing source rather than a case of one-off rapid destruction.

Repeating radio bursts include those labelled FRB 121002 and FRB 121102. Each of these codes is simply the date on which the burst first happened, with the year given at the start; so these two both occurred in 2012. The first-discovered FRB is labelled FRB 010724 because, although it was discovered in 2007, the discovery was made on the basis of going through archived data from the year 2001. While repeating bursts are clearly of potential interest to SETI practitioners, it seems more likely that they have a source that is unrelated to life than that the source is an alien intelligence – though the latter hypothesis has of course also been proposed.

## Morse and Arecibo

The first ever intentional human radio broadcast into space was in 1962, as we noted in the previous chapter. It was sent from an astronomical station at Yevpatoria, in the Crimea. This area was at the time part of the USSR. It became part of Ukraine in 1991; and it became Russian again through military intervention in 2014, though its status is currently a matter of dispute among the international community. The astronomical station has survived these political changes – though of course the radio telescopes and other equipment have evolved over time. The telescope that sent the pioneering Morse message has been replaced with an RT-70 model, although this itself is now in need of further upgrades, as it dates from the 1970s.

As its name suggests, the Morse message was sent in the form of Morse code. Strangely, I have a certificate of proficiency in this, dating from the 1960s when, as a schoolboy, I was a member of the Army Cadet Force. It's not something that's of much use these days. The code was devised by the American inventor Samuel Morse in the mid-nineteenth century, and had become widely used by 1900. Morse code was much used in both world wars; but since the digital revolution its usage has fallen significantly. It ceased to be the standard form of international distress signal for shipping in 1999.

The Morse message sent from Yevpatoria in 1962 consisted of three words: Mir, Lenin, and SSSR (in the Cyrillic alphabet, CCCP – the Russian for USSR). Taking Mir as an example, in Morse code it is two dashes (m), followed by two dots (i), and ending with dot-dash-dot (r). Of course, in 1962 the Mir space station (launched in 1986) was just a future possibility. But the word Mir, in Russian, means peace (and also, confusingly, world). This seems an appropriate first word to be sent into space (its transmission preceded that of the other two words by a few days). However, as might now be expected, the Russians received no answer.

The transmission of an ambiguous word by Soviet scientists, using an American-invented code, from a region that has been the subject of historically disputed ownership, within a country (USSR) that no longer exists, seems to me a fitting example of the combined hope, muddle, and conflict that is humanity. Our job in the twenty-first century as enlightened citizens of planet Earth is to try to extend the hope and to decrease both the muddle and the conflict. Easier said than done, of course.

Fast forward by a bit more than a decade and we get to the famous Arecibo message of 1974, broadcast by the radio telescope of the same name, situated in the municipality of the same name, on the north coast of the island of Puerto Rico. The message was devised by Frank Drake, Carl Sagan, and their colleagues. Although it was really only a symbolic message to demonstrate what the Arecibo Observatory – then recently upgraded – was capable of, rather than a real attempt to communicate with aliens (for reasons that we'll discuss shortly), it was a remarkable piece of work, and is worth examining in some detail.

The message was transmitted over a period of just a few minutes. It took the form of a stream of binary digits, but was intended to be read as a two-dimensional array of these on/off symbols. The stream was 1679 bits long. This number is the product of two prime numbers, 23 and 73. These give the dimensions of the array – 23 columns × 73 rows. The top rows indicated the numbers 1 to 10; then came information on the atomic numbers of H, C, N, O, and P – the main elemental constituents of life. This was followed by information on DNA, on humans, and on the solar system (see Figure 17). The 'third stone from the Sun' (Jimi Hendrix's famous term for the Earth) was shown displaced towards the figure of a human compared to the other 'stones', to indicate that we lived there. Interestingly, the number of planets shown was nine, because Pluto had not yet been demoted. The message ended with some information about the telescope that sent it.

Why was this superb piece of coding only of a symbolic nature? To answer this question we need to look at where it was sent: the globular cluster of stars called Messier 13 in the constellation of Hercules – a very beautiful object to look at with a small telescope. There are several reasons why no-one seriously trying to make contact with aliens would choose such a target. First, globular clusters are not thought to be particularly suited to life, since their stars are packed rather close together. Second, M13 is about 23,000 light years away, which means that the earliest we could expect to receive a reply would be 46,000 years

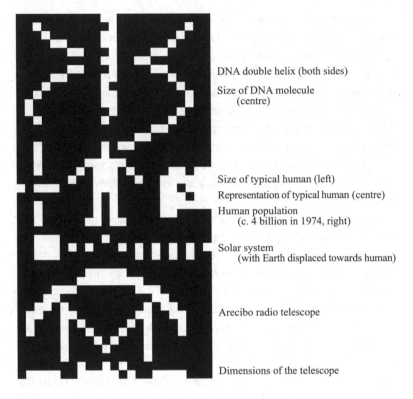

DNA double helix (both sides)

Size of DNA molecule
(centre)

Size of typical human (left)

Representation of typical human (centre)

Human population
(c. 4 billion in 1974, right)

Solar system
(with Earth displaced towards human)

Arecibo radio telescope

Dimensions of the telescope

**Figure 17** The part of the Arecibo message that includes the most visually interpretable information, which is (from top down): the double-helical structure of DNA; the approximate shape of a human; the solar system consisting of Sun and planets; and a graphic of the Arecibo radio telescope. All the other information shown simply consists of numbers (see text), given in binary form.

from now. Third, when the message gets to the position where M13 is now, the cluster will have moved elsewhere.

## Subsequent Outgoings

Since 1974, other messages have been sent into space. There haven't been many of them, especially considering how many years have elapsed, and some of them are hardly to be taken as serious contributions to SETI – for example the transmission of the Beatles song *Across the Universe*, which was sent from the Spanish hub of NASA's Deep Space Network (DSN) in 2008. The message was sent towards Polaris (alias the North Star or Pole Star), which is about 430 light years away. If we receive a musical message about 860 years from now, coming from the direction of Polaris, we'll know why. But the odds are against it.

*Across the Universe* was not the first musical message to be sent into space. It was preceded by *Teen Age Message*, sent from Yevpatoria in 2001. This message featured music from the electronic instrument called a theremin, along with other, non-musical, information. It was sent to five Sun-like stars, all within a distance of 100 light years. As well as music, poetry and environmentalism featured in the structure of the message. A more recent musical message sent in 2017 (and again in 2018) from a transmitter in Norway was directed towards Luyten's Star, which is a mere 12 light years away, and has a super-Earth in its habitable zone. As well as coded musical compositions, the message also included a tutorial on how to decode them.

A 2016 message with the rather odd name of *A Simple Response to an Elemental Message* (or ASREM) was broadcast to Polaris from the European Space Agency's Cebreros station in Spain, which is west of Madrid and quite close to the NASA Spanish DSN centre that broadcast *Across the Universe*. ASREM was also sent to Polaris, apparently due to this star's cultural rather than scientific significance. It's important to realize that this message was mistitled; it should really end with the word 'question' instead of 'message' because it wasn't a response to an alien message (it would be a lot more famous if it was), but rather to a human question, namely: how will our present environmental interactions shape our future? And it certainly wasn't a simple response, but rather a complex one. It consisted of more than 3000 separate responses from people in more than 100 countries, and it included answers in poetry as well as the more common ones in prose.

This leaves us with very few serious attempts to contact alien intelligence. *Cosmic Call* is the name given to two messages, both sent from Yevpatoria, but by a group of Texas-based business-men, in collaboration with scientists from several countries. Both involved complex message sequences, one part of which was the 1974 Arecibo message. *Cosmic Call 1* was sent in 1999 to four stars, all of which were at a distance of less than 100 light years. *Cosmic Call 2* was sent in 2003 to five stars, again less than 100 light years away. The two lists do not overlap, so the 1999 and 2003 calls covered a total of nine stars. The messages were sent at a frequency of 5 GHz (a wavelength of 6 cm). In theory, replies could start to come in after $2n$ years, where $n$ is the number of light years away for the closest of the nine stars; $n$ is about 35, so the minimum time-to-receive-reply is 70 years.

Two messages were sent, during 2008–09, to the star Gliese 581. These were entitled *A Message from Earth* and *Hello from Earth*. The former was sent from Yevpatoria, the latter from the Canberra Deep Space Communication Complex in Australia. Both broadcasts included diverse messages involving many con-tributors. Those of *A Message from Earth* were chosen via a com-petition on the social networking site Bebo. The Gliese 581 planetary system consists of at least three planets and pos-sibly five or more, including one – Gliese 581 d – that may be in the habitable zone (though its existence was still disputed at the time of writing). The distance to this system is about 20 light years, which means that it's one of the closest such systems to us. Given that these messages were sent in 2008–09, the earliest year for a reply would be 2048.

A reply to the Wow signal was sent in 2012 towards its apparent source. The *Lone Signal* was sent in 2013 towards the star Gliese 526, which is even closer than Gliese 581, at a dis-tance of about 18 light years. The former message, sent from Arecibo, is perhaps overly optimistic given that the Wow mes-sage was probably not from a biological source. The latter, sent from the Jamesburg Earth Station in Carmel, California, was a crowdfunded project; it involved a collection of messages from many people. Perhaps it also was on the optimistic side because (a) no planets are yet known to be orbiting this nearby star, and

(b) it's a flare star, so the habitability of any such planets is questionable.

Which of all the messages sent out so far should be taken seriously? Most can be excluded from consideration here because of their choice of targets. Those that might have a reasonable chance of being received by a nearby civilization that would be able to send us a reply that would reach us within a century are: *Cosmic Calls 1* and *2*; the two messages beamed towards the Gliese 581 system; and the two messages sent towards Luyten's Star. Perhaps *Lone Signal* is in there too, as we don't know for sure that a flare star precludes life. Even including it, that's just seven credible transmissions over the 46 years since the famous Arecibo message.

## Breakthrough

In 2015, the Breakthrough Initiatives were announced at the Royal Society in London. These initiatives, of which there are four (all related to SETI), are funded by the California-based, physics-trained, Russian businessman Yuri Milner. Also involved in the announcement of the initiatives were Stephen Hawking and Frank Drake. Milner's funding for the Breakthrough Initiatives runs to $100 million, which means that the scale of SETI activities will increase significantly in the decade from 2015 to 2025 compared to what it was up to 2015. The four initiatives are: Breakthrough Listen, Breakthrough Message, Breakthrough Watch, and Breakthrough Starshot.

Breakthrough Listen is the most advanced of the four, in that it has got to the stage of delivering results, albeit essentially negative ones so far. This initiative involves two radio telescopes – Green Bank, West Virginia, and Parkes, in New South Wales – and one optical telescope – the Automated Planet Finder (APF) at the Lick Observatory in California. The APF is searching for laser rather than radio signals – this endeavour is called Optical SETI. Overall, Breakthrough Listen aims to search for extraterrestrial signals from a million stars (and hence up to a million planetary systems). It got going in early 2016 and published its first set of results in 2017. Further results are expected to be announced once or twice a year. The initiative is coordinated from the Berkeley SETI Research Center, California.

Breakthrough Message is concerned with outgoing messages rather than incoming ones. However, instead of being a plan to send such messages imminently, it takes the form of a million-dollar competition to design such messages, together with a commitment not to send any message before there has been international discussion of the potential risks of doing so. Such risks have been raised before and to some extent ignored (see next section), but a revisiting of this issue is clearly timely if a major new round of 'active' SETI is about to begin due to the Breakthrough funding.

Breakthrough Watch is essentially a boost in funding for the search for quasi-Earths in planetary systems that are closer than 20 light years away. It will also include funding for analysis of the atmospheres of these planets. It has a particular focus on the nearest star system – Alpha/Proxima Centauri, at just over 4 light years – and in this respect it is linked to the Starshot initiative.

Breakthrough Starshot is a project to send tiny spacecraft to this system. The basic ideas are described in a paper entitled 'A roadmap to interstellar flight' by American scientist Philip Lubin. The spacecraft envisaged to travel to Alpha Centauri would weigh just a few grams, be powered by a system of ground-based lasers aimed at their light sails, and travel at 20% of the speed of light. They would be capable of making the journey in about 20 years. They would carry micro-cameras and other scaled-down equipment, and would be able to send information back to Earth in the form of radio signals that would reach us in just over 4 years from being transmitted, given the distance. So, allowing for a period of some years in development, we might be receiving information gleaned, for example, from photographing the planet Proxima b, in as little as 30 years. That's an incredible thought. As Lubin says: 'Exploring the nearest stars and exo-planets would be a profound voyage for humanity.'

As well as the four Breakthrough Initiatives discussed above, there are also Breakthrough prizes in mathematics, physics, and life sciences, each of which has now been awarded several times. I was particularly pleased to see a Breakthrough physics prize being awarded, in 2018, to Jocelyn Bell Burnell, who, during her PhD at Cambridge in 1967, discovered the regular pulses of radio waves initially dubbed the LGM signal (for Little Green Men) that turned out to be the first-observed pulsar (a rapidly rotating neutron star). The notorious decision of the Nobel Physics Prize Committee to award a prize to her PhD supervisor Antony Hewish but not to her in 1974 is to some extent compensated for by the award of her 2018 physics prize by the Breakthrough Foundation.

## The Opposite of Camouflage

Back in 1974, when Frank Drake was planning to send the Arecibo message into space, the UK's Astronomer Royal, Martin Ryle, contacted him and urged caution on the basis that we might not want to advertise our existence to intelligent extraterrestrials, given that they might be more advanced than us and might be potentially hostile. Drake's response was that he didn't really see any problem because we'd been leaking radio waves into space ever since the start of regular broadcasting in the 1920s. Given this stance, he went ahead and sent the message.

The idea that deliberate messaging to space carries with it no more risk than the accidental escape of commercial radio broadcasts is fundamentally wrong. The reason for this has to do with the power of the signals. It's true that aliens training radio detectors towards Earth might detect a daily pattern in the amount of radio they were receiving, given that peaks of population such as western Europe and the eastern and western seaboards of North America will show up as peaks at certain times of day. This is because leakage has a directionality – terrestrial broadcasts are sent across the Earth and so will leak out mostly at a tangent to the Earth's surface. An alien listening to us from a nearby exoplanet in a particular direction will perceive at least three peaks per day, perhaps with a fourth representing Japan.

But the key phrase here is 'from a nearby exoplanet', because radio leakage is relatively weak. In contrast, a powerful signal beamed out intentionally towards a particular exoplanet will be much stronger, and so has the potential to be received and understood much further away. Given our uncertainty about the number of broadcasting civilizations, and hence the distance to the nearest one, the difference between the maximum travel of leakage and that of deliberate interstellar signals might correspond to the realm within which the nearest such civilization exists.

We should add to this concern the fact that interstellar signals like Arecibo are designed to be seen/heard and understood. In this respect they represent the opposite of camouflage, where the aim is to remain unseen – or seen but understood to be something else, as in the case of making a military vehicle look like vegetation to enemy aircraft. This kind of camouflage is appropriate in warfare but not needed in peacetime. The trouble is that with as-yet unencountered aliens, we don't know whether to expect war or peace. Thus we don't know whether sending out signals whose content is designed to be interpreted by a technologically advanced civilization is a very good idea or a very bad one.

In this context, the stimulation of debate by Breakthrough Message is sensible and timely. Although its outcome cannot be anticipated in detail, we should expect that as in any other surveys of human opinion there will be a diversity of views. What would the best response be to such a diversity? Perhaps go with the majority in terms of a decision to send or not to send? Alternatively, perhaps listen to the most informed views? There's no easy answer. But here's my own view, for what it's worth. The first members of our species to leave Africa about 100,000 years ago did not fear the unknown sufficiently to deter their travels. And European explorers of the fifteenth century sailed off into the unknown with death or discovery being about equally likely. Many died, but some discovered new continents. We are a curiosity-driven species; science is a curiosity-driven venture. We will never stop being curious and seeking new knowledge unless or until it kills us. It's better to risk an explorer's death than to hide in the darkness of ignorance.

*Part V*

# Beyond the Milky Way

## Key Hypotheses

### The Huge Hypothesis
The biological universe was born more than 10 billion years ago. The vast majority of the life that comprises it is carbon-based; most but not all life-forms are constructed on a cellular basis; on many planets large (multicellular) organisms have evolved; in a tiny proportion of these organisms, intelligence has evolved.

### The Trillion Intelligences Hypothesis
Although radio-level intelligence is very rare in proportional terms – perhaps as rare as one civilization per galaxy – there are probably more than a trillion intelligent civilizations in the observable universe right now, most of which are more advanced than we are.

# 18 THE PHYSICAL UNIVERSE

## Photograph Extraordinaire

We live in privileged times. Anyone with access to a computer has, just a few clicks away, the Hubble Ultra-Deep Field (HUDF) – a beautiful image of myriad galaxies seemingly stretching out to infinity and beyond. Anthony Doerr, author of the novel *All the Light We Cannot See*, described it in 2007 as 'the most incredible photograph ever taken'. I believe he's right in this, and also in saying that 'The Hubble Ultra Deep Field image should be in every classroom in the world.'

Let's look at this image (Figure 18) as closely as its reproduction in black-and-white on a smallish page permits – I would strongly urge readers who haven't seen it before to type HUDF into Google Images to see it in its true splendour. We'll start with how it was obtained – an almost miraculous feat of precision technology. The HUDF is a photo of a small portion of sky in the direction of the southern constellation of Fornax (the furnace). 'Small' here means about one thirteen-millionth. So, with a few ifs and buts that we'll get to shortly, whatever we see there needs to be multiplied by 13 million to get the whole picture. The photo was obtained by pointing the Hubble Space Telescope at this particular tiny patch of sky for a period of about 11 days. In photographic terms, that's the exposure time. However, this phrase doesn't mean quite the same thing for an orbiting space telescope as it does for a hand-held camera. Rather, the 11 days' worth of exposure was obtained in bits over a period from late 2003 to early 2004.

**Figure 18** The Hubble Ultra-Deep Field photograph, which includes only a handful of Milky Way stars in the foreground (distinguished by a cross-like pattern, which is a telescopic artefact) but thousands of distant galaxies in the background (everything in the picture that lacks a cross is a galaxy). There is one particularly clear spiral galaxy near the middle of the lower-right quadrant of the photograph. Courtesy of NASA.

The area of the sky photographed was chosen to have as few Milky Way stars as possible in the foreground, since these occlude what's behind them. In the end, there were a mere six in the picture. Contrast that with the number of galaxies: around 10,000 in the high-resolution version, though far fewer are visible in a reproduction such as that shown here. There are some galaxies that are clearly spirals, like our own. There are others that look like bright dashes – these are probably spirals seen edge-on. And there are others again that are fuzzy blobs – probably a mixture of elliptical and irregular galaxies. Some of them are so far away that they just look like points of light, and yet each point is the combined light of many billions of stars. Although each galaxy in which a spiral structure is visible has a plane of symmetry, like the Milky Way, the overall picture shows that the various galactic orientations are all higgledy-piggledy; for example, a face-on spiral and an edge-on spiral can be seen apparently jostling up against each other – though we always need to remember that invisible third dimension of the sky.

More than any other single image, this photograph reveals the vastness of our universe. Our own galaxy shrinks into insignificance. This is like Carl Sagan's famous 'pale blue dot' (an image of Earth taken by *Voyager 1*) writ large. From the outer reaches of our solar system, the Earth is indeed just a dot. But from the perspective of an inhabitant of one of the most distant galaxies in the HUDF, our entire home galaxy would itself be a dot – assuming they have taken an equivalent of the Hubble image, with their space telescope pointing in our direction. We've come a long way from the days of the Earth apparently being at the centre of things (pre-Copernicus) and even from the more recent days (early twentieth century) of some scientists arguing that the Milky Way and the universe were one and the same thing.

## The Observable Universe

The Ultra-Deep Field suggests immeasurable distances. But measurement on the scale of the utterly vast is not entirely impossible. We have a range of astronomical measurement techniques at our disposal, some of which are particularly useful at intergalactic scales, notably measurement of a galaxy's redshift. The greater the redshift, the further away is the galaxy, with Hubble's law providing a reasonable quantification of this relationship, at least for galaxies that are only a few millions of light years away. Beyond that spatial scale, the proportionality between redshift and distance breaks down due to the accelerating expansion of the universe, but the general principle of 'the higher the redshift the greater the distance' remains.

The most distant galaxy that we've seen so far is about 13.4 billion light years away. But how much further than that might it be possible to see? This is a tricky question. Let's invent a hybrid universe that incorporates elements of the old idea of a static model and the newer Big Bang model. Let's think of a universe that came into being almost instantaneously about 13.8 billion years ago (bang) but when it appeared had much the same structure at an intergalactic level as it does now, and hence has looked pretty much the same ever since (stasis), and hasn't expanded after the initial blast. In such a universe, the furthest we could see would be 13.8 billion years. All galaxies further away than that, of which there might be many, would be invisible to us today because the light from them would not have had time to reach us.

However, the actual universe is not like that, as we now know. Given that it started very small and expanded at a variable rate, we should be able to see galaxies that are further away than 13.8 billion light years in terms of distances in present-day space. The current record-holder that is 'only' 13.4 billion light years away in what's called 'light travel time' (the galaxy GN-z11) is more like 32 billion light years away within the spatial framework of the present universe. In other words, the light we are now seeing from it has travelled only a fraction (about 40%) of the current intergalactic distance, because when it started out that galaxy and ours were much closer together.

It has been calculated that the furthest we could ever see in terms of present-day distances is about 46.5 billion light years. If this is correct, then the diameter of the observable universe is twice that – about 93 billion light years. This is an easy number to remember if your brain uses miles instead of (or as well as) kilometres to measure distances, because the figure 93 also occurs in the familiar distance from the Earth to the Sun (93 million miles). It's interesting to ask how many times bigger the first of these figures is than the second. Since a light year is about 6 trillion miles, the answer is approximately 6 quadrillion times (with a quadrillion being $10^{15}$).

Thus we can expect that as telescope technology gets ever better we will be able to see ever further than is currently possible – but only up to a certain point. We will eventually hit a brick wall. It seems unlikely that there's an actual brick wall around the observable universe; but in terms of seeing things there might as well be. Then again, we can't complain too much, because the limiting factor right now is our limited technology, not (yet) the limited speed of light.

The number of galaxies in the observable universe is currently thought to be about two trillion. This means that if we want to estimate the number of occurrences of 'something' in the observable universe from measurements made within the Milky Way, we need to multiply by this number – though there are a few provisos to such a multiplication. The most obvious one is that we are tacitly assuming that the Milky Way is a typical galaxy, which it's probably not. It's perhaps close to being a typical spiral galaxy, and closer still to being a typical barred spiral. So the size of the optimal multiple depends on what the 'something' is. If it's a galactic centre, then two trillion is probably fine, since every galaxy has a centre, even though in some cases – for example irregular galaxies – we may be stretching the use of 'centre' somewhat. But if the 'something' is life, and if life is only found on planets within spiral galaxies (unlikely), then our multiple would need to be smaller by a factor that reflects the proportion of all galaxies that are spiral, which is thought to be about 50%; so we would multiply by one trillion rather than two, to get from the Milky Way to the observable universe.

But wait a minute. Our knowledge of the proportions of different types of galaxy is intrinsically biased by the sample of them that we've seen, and this sample is in turn biased by location in space – we know the most about the closest galaxies to us. So we have to ask another key question at this point: is our home location in the universe typical or in some way special? The short answer is 'typical', but we should delve into this issue, which in fact turns out to be two rather separate issues. Each of them is the subject of an important principle, as we'll now see.

## The Cosmological Principle

A casual version of this principle is: one part of the universe is pretty much the same as another. But although these days most scientists favour casual dress, they certainly don't favour being casual with proposed generalizations. After all, generality of explanation is at the heart of science. So we shouldn't be content with the casual version of the cosmological principle. Here's the more formal version: at sufficiently large spatial scales, the universe is both homogeneous and isotropic. As can be seen, the formal version differs from its casual counterpart in two ways: first, it restricts the principle to high levels of scale; and second, it separates out two ways in which things might or might not be 'the same'. Let's scrutinize these in turn.

The restriction in scale is obvious enough. At small levels of spatial scale different parts of the universe are manifestly not even nearly the same. Indeed, the whole concept of the habitable zone is based on this. Near to a star, further away from it, and much further away again, are all very different places, as attested to by a comparison of Mercury, Earth, and Neptune. Going up a level, the centre of the Milky Way, its periphery, and a point halfway between it and Andromeda, are again very different places. Going up another couple of levels, our galaxy supercluster (Laniakea) has a different size and shape from other neighbouring superclusters; and all of them are very different from points in inter-supercluster space.

How far up do we have to go to obtain homogeneity of different bits of the universe? Estimates span two orders of magnitude: from 100 million light years to 1 billion light years. Taking one random billion-light-year cube of universe and comparing it with another, we should find that to all intents and purposes they are the same. However, whether this is true or not is a moot point. We keep discovering bigger and bigger structures at the supergalactic level. These discoveries strain the cosmological principle. One of the biggest structures in the known universe is the Sloan Great Wall, which is more than 1 billion light years across. It would be easy to find one big cubic sample that contains part of such a wall, and another that doesn't, so we might have to revisit the minimum scale limitation of the principle. But that wouldn't fundamentally change it.

Now we get to the separating out of those two aspects of sameness – homogeneity and isotropy (the corresponding aspects of otherness being heterogeneity and anisotropy). Let's start with homogeneity.

Imagine a small flat metal grid with 10 rows and 10 columns, each separated from the next ones by metal bars, and a square metal plate providing a base for the overall grid. Each of the 100 holes is big enough to hold a ceramic ball that's about the size of a pea. If all 100 holes are occupied by balls that are identical, we have a homogeneous array; if the balls are different, then our array is heterogeneous. At this level, homo- versus hetero-geneity is acting as a binary variable. But there's no need to stop at that point, because we can recognize different levels of heterogeneity. For a low level, the balls might be just black or white. But for a higher level they might consist of many colours. And for a higher level still they might also vary in surface texture – for example, smooth versus rough. And so on. Essentially, homogeneity is a single state, while heterogeneity is an infinite series of states.

Now let's return from our small metal grid to the universe. At low levels of spatial scale, the universe is heterogeneous, because the 'balls', whether planets, stars, galaxies, or clusters, are all different from each other. But at high levels of spatial scale it's homogeneous, at least according to the cosmological principle, because one huge collection of these things is essentially the same as another. Note the 'essentially' qualification that has crept in here. Planets are not mass-produced like ceramic balls; nor are stars, though as we saw earlier, their production is more law-like and less accident-prone than that of the planetary systems that surround them. Nor are galaxies identical to each other. But none of those things stop big chunks of the universe being essentially identical; hence they don't stop the universe from being essentially homogeneous at a large scale.

We should now consider the other claim of the cosmological principle – that the universe is isotropic. To do this, we should start by asking what isotropy is, and how it differs from homogeneity. A system (whether of ceramic balls or of galaxies) is said to be isotropic if it is the same in all directions. You might think that a homogeneous system, as discussed above, is necessarily also an isotropic one – but you'd be wrong. A homogeneous system can be either isotropic or anisotropic. To get an idea of how the latter (counterintuitive) combination might arise, consider the possibility that the entire universe, like the Earth and the Sun, has a magnetic field. In this case one direction across the universe is north, and opposite that is south. Not only are these different from each other, but the north–south direction of the universe has a different nature from the east–west direction. However, these specifics of directionality don't really matter, because the issue is not whether the universe has a giant magnetic field (it doesn't), but rather whether it might be suffused by *any* type of 'field', including types we don't yet know about, that have directionality. According to the cosmological principle, it doesn't. In other words, at sufficiently large spatial scales, the universe is not just homogeneous, but isotropic too.

## The Copernican Principle

I deliberately didn't refer to an observer in the above account of the cosmological principle, because taking an observer-free approach helps to emphasize the difference between this principle and its Copernican counterpart. Perhaps an observer was implicit in the last section, because we were doing things like comparing two giant cubic samples of the universe; hence 'we' were observers who could in theory see, or at least envisage, these samples. But let's eliminate such implicit observers by imagining them as residing in a different universe and having a magic window into ours. (That in itself would render ours heterogeneous, but let's try to stay sane here and ignore that awkward fact.)

The Copernican principle applies not so much to the structure of the universe as to our location, as human observers, within it. And one casual version of it is: our location is not special. However, a problem with the Copernican principle is that there isn't really an agreed formal version, so all we can do is to look at different ways of stating it and to see what they have in common. Other ways of stating the Copernican principle are as follows: our location is not central; our location is not privileged; it is ordinary; it is mediocre; it is humdrum.

The only one of these versions of the principle that is more specific than the others is 'we are not central'. And this version connects well with what is often called the Copernican revolution – the acceptance that the Earth is not at the centre of the solar system, as was originally thought by most astronomers before Copernicus's *On the Revolutions of the Celestial Spheres* (published in Latin in the year of his death, 1543). Perhaps the use of 'solar system' here is questionable, as some of the planets were unknown in the 1500s (Uranus and Neptune), and the huge gulf in scale between the solar system and the galaxy was not fully comprehended, even though the planets were known to behave differently to the stars. The even huger gulf in scale between the galaxy and the universe would not be known for another four centuries.

However, whether we take the Copernican principle to state that we're not at the centre of the solar system, or at the centre of the Milky Way, or at the centre of some larger structure, for example our galaxy supercluster, it's true. Strangely, whether it's true or not at the level of the universe depends on what we mean by 'universe'. We *are* at the centre of the observable universe, but then again a civilization living on a planet in another galaxy will also be at the centre of its own observable universe. And that's simply because an observable universe is a concept that is defined from the reference point of a particular place. There is no one absolute observable universe, but rather countless relative ones.

Might we be at the centre of the universe itself? In other words, summing both the observable universe and whatever lies beyond it, might we truly be at the centre of everything? It's tempting to say 'no' simply on the basis of countering the curse of human arrogance, something that we keep fighting against with varying degrees of success. But the real answer is harder to come by. If the universe is infinite in space, then whether or not it has a centre point is perhaps a matter for philosophers rather than scientists. If it's merely ultra-huge but finite, then the probability of us being at the centre is vanishingly low.

So, the version of the Copernican principle that claims we're not at the centre of things is correct, with the trivial exception of a 'thing' that's defined by our location in the first place, namely the observable universe. But there's a lot in a single word. Change 'central' to 'special' and the (revised) Copernican principle is manifestly wrong. There's probably nothing terribly special about our Sun, especially when compared to other G-class stars in the Milky Way's disc. And there may be nothing special about this disc compared with its counterparts in other spiral galaxies. But there is something *very* special about habitable zones, wherever in the universe they exist: the possibility for a planet to have liquid water on its surface in the form of lakes or oceans. In this respect, our location *is* special, but so too is that of every other biosphere in the universe. So, compared to them, we're back to being 'unspecial' again.

## Seeing Specific Events

The beautiful Hubble photo with which this chapter started was a landmark in our visualization of the universe. It has been enhanced since 2003–04 when it was taken, by the addition of ultraviolet and infrared wavelengths. And no doubt in the future our ability to photograph distant galaxies will get even better. But in the end all such photos are essentially static. They appear to show motionless galaxies occupying fixed positions in the vastness of space. There's a clue that not all is stasis, because the spiral structure of some of the distant galaxies suggests long-term rotation of the sort that we know to characterize the Milky Way. But that's about it.

What do we need to do in order to see not just other galaxies but particular cosmic events? At first it seems like a hopeless case. Taking a video rather than a photo is a possibility, but for motion as slow as galactic rotation – recall that the solar system takes about 250 million years to rotate once around the Milky Way – a video is not much use either. However, some things in the universe happen quickly – not just in astronomical terms but in human terms too. Some cataclysmic events in space occur on timescales of seconds, minutes, hours, days, or weeks, rather than millennia and beyond. A good example of a major but short-lived astronomical event is a supernova explosion, which appears as a spike of dramatically increased light typically lasting a few weeks, visible to the naked eye in a few cases and through telescopes in others.

Supernovae have been observed since antiquity. We know this because of historical records, and because some of these records are good enough to allow us to connect a current supernova remnant with a supernova explosion of the distant past. A good example of this is the Crab Nebula, visible to amateur astronomers as a pale grey luminous patch in the constellation of Taurus. This nebula is what is left after a supernova explosion in the year 1054, which was recorded in detail by Chinese astronomers. We now refer to that ancient supernova as SN 1054. And in general, supernovae are coded in the form of SN followed by the year in which they occurred; in many cases, additional letters are used to distinguish supernovae that occurred in the same year or to provide other information.

More recently observed supernovae include SN 1987A and SN 2014J. These two illustrate the different levels of spatial scale on which we can see these amazing things. The former involved the death of a star in the Large Magellanic Cloud, one of the dwarf galaxies that are gravitationally bound to the Milky Way. This event thus occurred less than 0.2 million light years away. The latter supernova occurred in the Cigar Galaxy – so called because we see it edge-on and so through a small telescope it looks like a distant celestial cigar. It's approximately 11 million light years from us. Photos taken before and after in these and other cases reveal that a supernova can often rival its host galaxy for brightness, with a single exploding star being very visible as a bright dot against the grey smudge of its many compatriot stars.

Of course, our knowledge of specific events in the physical universe is not restricted to supernovae. Some other phenomena with very short timescales that we've 'seen' (in various ways) include pulsar rotations and collisions between neutron stars or black holes.

Pulsar rotation has been detected by periodic spikes of radiation with wavelengths in the gamma-ray, x-ray, and radio zones of the electromagnetic spectrum. Collisions between massive objects such as neutron stars and black holes have recently been observed using gravitational wave detectors – notably LIGO (the Laser Interferometer Gravitational-Wave Observatory) in Washington State and Louisiana, and the Virgo interferometer in Tuscany, Italy. These waves travel at the speed of light, but the distortions of spacetime that they represent are so tiny that distinguishing a real signal from noise is fraught with difficulty. The first detection of gravitational waves – deriving from a collision of black holes – was in 2015. And the first time that an event was observed to trigger both gravitational waves and bursts of radiation – a collision of neutron stars – was in 2017. This event was of huge importance: it helped to validate the gravitational wave technique. The 2017 collision happened in a galaxy more than 100 million light years away. However, that's close compared to the first gravitational wave detection, which resulted from a collision at a distance of more than a billion light years.

We saw earlier that fast radio bursts (FRBs) come from distant galaxies. Now we know that the events generating gravitational waves have a similar origin in terms of distance. And supernova explosions that we observe are typically beyond the Milky Way – though if the red supergiant Betelgeuse in the constellation of Orion goes supernova soon, that statement will seem a bit odd. Supernova explosions are seen through optical telescopes, but they are often accompanied by bursts of gamma rays. So, overall we have evidence of extragalactic events from many parts of the electromagnetic spectrum and, since 2015, from gravitational waves too.

## Low-Power Life?

It's clear from the above that the physical universe reveals itself in many ways. But what about the biological universe (aside from Earth)? This putative entity, the nature of which we'll contemplate in the next chapter, has been remarkably quiet so far. There are several possible reasons for this apparent bio-silence. One is that – apart from us – it doesn't exist. But the idea that Earth is the only planet with life in more than two trillion galaxies is preposterous. An alternative possibility is that the apparent bio-silence of the universe is to do with the relatively low power of signals that indicate life, compared with the immense power of signals stemming from events in the non-biological universe. Let's explore this possibility, in relation to intelligent life in particular.

The silence of intelligent alien life – to date – takes us immediately to the Fermi paradox, named after the great Italian-American physicist Enrico Fermi. He famously asked the question: if there are lots of intelligent civilizations out there, how come we haven't heard from any of them? There have been many proposed resolutions of the Fermi paradox, one of which is that the closest civilizations are very distant (perhaps outside the Milky Way) and their radio signals are not powerful enough for us to detect.

To consider this possibility, we should start by thinking about how power is measured. A commonly used unit is the watt (W), named after the Scottish scientist James Watt (1736–1819), who invented the Watt steam engine – a device that played an important role in the Industrial Revolution. We're familiar with watts because the power of domestic appliances is often measured using these units. For example, the cooking instructions for microwave ovens usually refer to various powers of up to about 1000 watts (1 kilowatt).

For radio communication within the solar system, those sorts of numbers are more than enough. *Voyager 1*, which is currently at the edge of our system, transmits messages back to Earth with a power of about 20 watts. However, when these reach Earth, their power has dropped to a tiny fraction of 1 watt. This is still receivable and interpretable, despite the low power. But if *Voyager 1* decided to change course and head for Alpha Centauri (unlikely for multiple reasons), then when it got there we would be unable to receive its messages.

Why does the signal lose power at all? For the vast majority of its journey, it's travelling through space, not air. And while we now know that space isn't quite empty, its density of particles is extremely low. The answer lies in something called the inverse square law, which sounds complicated, but isn't. Like the beam from a flashlight, a radio signal spreads out as it travels, which means that its power is diffused over a progressively wider area. Thus the proportion of the beam going in the intended direction gets smaller and smaller with distance. In fact, it gets smaller in proportion to the square of the distance – hence the law's name.

Now compare the 20-watt signals from *Voyager 1* with the Arecibo message. This was very much channelled towards its target (a globular star cluster in the Milky Way, as we saw earlier), but it was the equivalent of an unchanelled signal sent at a power of about 20 trillion watts. It's thought that this would be receivable and interpretable across most of the galaxy. This shows that human life is not necessarily so low-powered after all. However, even the high-powered Arecibo message would probably not survive intergalactic travel well enough to be interpretable to a civilization outside the Milky Way, when it eventually reaches one in millions of years. And the power of this message pales into insignificance compared to that of beams of electromagnetic radiation from physical sources. The power radiated by the Sun is $3.8 \times 10^{26}$ watts (or $3.8 \times 100$ septillion watts, if you prefer). Other high-power physical sources often have their output measured in multiples of this unit. For example, Betelgeuse has a power approximately 125,000 times that of the Sun.

How do considerations of power apply to our ability to detect non-intelligent life-forms? As we noted in Chapter 16, in the section entitled *Imagining Radio Eyes*, organisms on Earth are unable to send or receive radio waves – in contrast to their ability to emit (bioluminescence) and detect (sight) visible light. So the power of radio signals has no relevance in relation to our ability to detect biospheres in which the most complex creature is a cat. However, power applies to signals that are sent and received at any wavelength. It thus applies to the visible and infrared light that we discussed in Chapter 15 in relation to biosignatures. So, as we now move on to explore the biological universe generally (Chapter 19) and the intelligent component of it specifically (Chapter 20), we should keep in mind that signal power is always an important practical consideration.

# 19 THE BIOLOGICAL UNIVERSE

## An Awe-Inspiring Thought

Think again of that amazing photograph of thousands of galaxies that we call the Hubble Ultra-Deep Field (Figure 18). Now imagine that all of the galaxies in the photo that harbour life on one or more of their many planets are coloured green (picture a bright green, like that of beech trees in spring), while the rest are left as they are, and thus appear white (or yellow or orange in the original version). What proportion would be green? Although we can't give a definite answer to this question, the notion that our own galaxy isn't special – part of the Copernican principle, as we saw in the previous chapter – suggests that our tentative answer should be 'most of them'. Whether 'most' should be 51%, 75%, or 99% is a moot point. But even if it's only the first of these, it's an awe-inspiring idea that in a single photograph we can perhaps see more than 5000 galaxies with life, in a field of view that, as you'll recall, represents only a tiny fraction of the sky.

From this amazing initial thought, we can go in two directions. First, we can select any one of the green galaxies and focus inwards. Imagine a blown-up photo of one particular spiral galaxy in which all the stars that have planetary systems with life are coloured green, the rest left a more natural white. As we've already seen from consideration of the Milky Way, the chances are that the number of green stars runs into the billions. Second, we can shrink the famous Hubble photo and imagine it as one piece of a patchwork of photos covering the whole sky. Although most of these photos don't yet exist, they probably will one day. And when they do exist, if we were to do the 'green-for-life' colouring-in act, we'd see green galaxies spread all over our array of photos – probably more than a trillion of them.

What we're approaching here is the extent of what I call the biological universe. This is the collection of all life-forms that exist right now – or, in a time-extended view, all life-forms that have ever existed plus all those that are yet to be born. Its current spatial extent is vast, like that of the physical universe in which it is embedded. Taking it backward in time, it doesn't have as long a history as its physical counterpart, but the difference might not be very great. Assuming that our current estimate of the age of the physical universe (13.8 billion years) is about right, the age of the biological universe is probably greater than 10 billion years and might even be 13 billion or so. One important limiting factor for life in the early universe was the low metallicity of the earliest planetary systems, as we saw earlier.

Going forward in time, we currently think that the future of the physical universe will be continued accelerating expansion leading eventually to a 'big freeze'. If this is correct, then the biological universe will extend less far into the future than its physical counterpart, because the final stages of the 'freeze' will be uninhabitable for a host of reasons, with low temperature being just one of them. So in this respect there is a temporal symmetry: the physical universe started before the biological one, and will still exist after the last life-form within it has died.

Thinking about the possible extent in space and time of the biological universe is hard enough; thinking about *what it might be like* either now or in the past or the future is harder still. Indeed, a reasonable view might be that since we currently only have a sample size of one known life-bearing planet, embedded within one known life-bearing galaxy, any attempt to think beyond that in terms of the types of life-form that make up the rest of the biological universe would be premature. However, an alternative view is that we Earthly life-forms are not special; we're more likely to represent a commonly recurring form of life in the universe than a vanishingly rare one. Not only is our celestial neighbourhood not special, but neither is our chemical composition. I call this stance Copernican biology, and I believe it has much to recommend it. It features a lot in the rest of this chapter. Indeed, it has featured a lot in all the preceding chapters, but only implicitly, as I have not used the phrase 'Copernican biology' until now.

Scaling up from life in the Milky Way to the biological universe is both harder and easier than scaling up from our solar system to the Milky Way. It's harder because we have no solid information on the existence of Earth-like planets in other galaxies. But it's easier for what we might call a Copernican reason. When we first delved into the Copernican principle that our location in space is not special, we came to the conclusion that this was true at some spatial scales but not at others. Our location in the solar system clearly is special – in the Sun's habitable zone. Our location in the Milky Way is much less so. However, we are in the galactic disc rather than in the bulge or the halo; and although there is no well-defined galactic habitable zone, it's possible that those other parts of the galaxy are on average less conducive to life than is the disc. If this is the case, then it might be more likely that there is an Earth-like planet in the disc of a billion-light-year-away spiral galaxy than that there is such a planet near the supermassive black hole at the centre of the Milky Way or in a globular cluster somewhere in the outer reaches of its stellar halo.

Which of the 'harder' and 'easier' features of scaling up from galaxy to universe will get the upper hand? This is where that other important generalization about the universe – the cosmological principle – comes in. We saw in Chapter 18 that under this principle the universe is homogeneous at large scales. This means that a spiral galaxy a billion light years away should have a similar array of planets to the Milky Way and its neighbours, such as Andromeda. If the cosmological principle is correct, then our lack of solid information about exoplanets in galaxies other than our own really isn't important. We can be reasonably sure that they're there, even if we have no direct evidence of their existence.

Let's briefly consider the alternative possibility – that the cosmological principle is wrong. In that case, there might be planet-less galaxies that in other respects resemble ours. These might include ones that we can see, for example in the Hubble Ultra-Deep Field. Some might be uncannily similar in shape and size to the Milky Way. Yet somehow all their constituent stars have formed in a different way than local stars, with no proto-planetary discs surrounding them to give birth to planets. This seems such a bizarre notion, implying an almost mischievous universe whose job is to trick us, that it seems reasonable to ignore it and instead to go along with the idea that, on a sufficiently large scale, the universe is indeed homogeneous.

## The Laws of Science

Let's be clear about one thing in relation to our discussion of the biological universe. It's a part of the physical universe, not something aside from it. In terms of set theory, their relationship is inclusion: the biological universe is a subset within the larger set that we call the physical universe. (The area within the larger set but outside its included subset can be called the non-biological universe.) Given this relationship, it's clear that all the life-forms that make up the biological universe must obey the laws of science, however far away from us those life-forms may be. Or, to qualify this statement, they must obey all *applicable* scientific laws. But which laws exactly would those be? Let's now explore this question.

Almost all scientists take for granted the universality of the laws of physics, such as $E = mc^2$. We do not expect these laws to be dependent on location in space or time. This is thought to be equally true of things that we're personally familiar with (like sunlight) and things that we're not (like black holes). The electromagnetic spectrum is no different in one galaxy than it is in another. Likewise, the supermassive black hole at the centre of the Milky Way is not fundamentally different from its counterpart at the centre of another galaxy of similar size and shape elsewhere in the universe.

Whether the same can be said about the laws of other sciences is less certain; hence the jokes about 'physics envy'. However, perhaps chemistry comes close. Indeed, perhaps it gives physics a run for its money in this respect. Would we expect chemical reactions to differ between one part of the universe and another? Probably not. Under certain conditions here on Earth, sodium and chorine will react to give salt (sodium chloride). Equally, under certain conditions, iron and oxygen will react to give rust (iron oxide). On the surface of a rocky planet in another galaxy, we would expect that the same reactions would take place, unless there is some 'unusual' difference between conditions here and conditions there. And even if there is, we would expect to be able to replicate the alien conditions in an Earth laboratory, and find that their effects on chemical reactions are the same here as they were on the 'unusual planet'.

But can we expect other sciences to rival physics in this respect too? Both biology and geology seem to have more of a local flavour than King Physics and Queen Chemistry. For geology, we now know enough about our solar system to be confident that some of Earth's geological features are unique at this level of spatial scale. We seem to be the only local planet that has plate tectonics. And the only solar system body other than Earth that has active volcanism (in the sense of 'normal' volcanoes rather than cryovolcanoes) is Jupiter's moon Io.

Other geological features are probably similar in general and just different in detail among the rocky planets. A good example of this is what's known as differentiation. In any solid body of sufficient size, whether a planet or a large asteroid or moon, the heaviest material sinks to the centre, leaving lighter material at the periphery; hence the Earth's core of iron and nickel and its lighter mantle and crust of silicates. The other local rocky planets also have metallic cores, though they occupy different proportions of each planet's interior. The Earth's core makes up 16% of its volume. The corresponding values for Mercury, Venus, and Mars are approximately 40%, 12%, and 9% respectively. So the same phenomenon of differentiation has occurred in all these planets, but the details differ somewhat. Perhaps an appropriate law for this would be: any solid body in space above a certain mass undergoes differentiation of its materials.

Life-forms don't have to obey this law, as they're not large bodies of rock in space; though their home planets *do* have to obey it, unless they constitute some sort of as-yet unknown exception. This fact illustrates the difference in applicability of the laws of physics on the one hand and the laws of geology and planetary science on the other. Physics encompasses life, while geology doesn't. Life-forms must obey (a) the laws of science that apply to the entire universe, and (b) those laws that apply just to its biological subset. They don't have to obey laws that apply only to the non-biological universe, or parts of it.

What about the laws of Earth-based biology? Is it reasonable to expect that the general features of life on Earth also characterize life elsewhere? This is the million-yen question. The first step in answering it is to acknowledge that biology is in a sense part of chemistry. The metabolism that powers life consists of a very complex network of reactions among molecules, and can be thought of as a branch of organic chemistry that operates within cells. Or, to put it another way, the rules of biochemical reactions are not fundamentally different from the rules of chemical reactions in general – biochemistry is simply one realm out of many for their application.

Biology is also part of physics. If we focus on energy rather than chemical reactions, it becomes clear that biological energy transfer systems must obey the physical laws that govern energy flow in a broader context. For example, energy flow in ecosystems has to obey the law of conservation of energy; the biological realm has no special privilege in this respect. Thus as energy flows from plants to herbivores to carnivores, each transfer involves loss of energy from a biological point of view, as we saw in Chapter 5. However, if we were to take into account also all the non-biological forms of energy, notably heat, and build a precise energy budget for the overall food chain (or web), we'd find that the amount of energy of all types added together remains constant.

Strangely, although biology doesn't escape the rigours of the first law of thermodynamics (conservation of energy), it does seem to escape the rigours of the second law, the law of increasing entropy or disorder. In particular, the short-term process of embryonic development and the long-term process of biological evolution clearly lead to more complex, ordered states on their respective timescales. Here, entropy is going down, not up. But of course the apparent breaking of the second law isn't real. As we saw in Chapter 5, the law only applies to a complete, closed system. The biosphere isn't such a system – rather, it's a steady-state system in which there is continual energy inflow and outflow. The second law doesn't preclude such a system from decreasing in entropy, so long as, within the larger system of which it is a part, any local decreases are more than offset by increases elsewhere.

So, on planet Earth, chemical reactions in biology obey the laws of chemical reactions in general. And the flow of energy and entropy in biology are constrained by the laws of thermodynamics just like such flows in the abiotic realm. Against this background, should we expect to find life elsewhere that is not constrained in these ways? The answer is a resounding 'no'. With our current sample size of only one known biosphere, it is indeed a good idea to keep an open mind about the nature of alien life. Then again, we should remember Carl Sagan's famous piece of advice: 'It's good to keep an open mind, but not so open your brains fall out.'

Now we return to the million-yen question: should we expect the laws of Earth-based biology to apply in other biospheres, whether in the Milky Way or beyond? Before taking the second step towards answering this question (recall that the first step was simply acknowledging that biology is part of chemistry), we need to reach a consensus on what the laws of terrestrial biology are. This isn't easy, because most of the 'laws' of biology have exceptions. Under a strict interpretation of 'law', they aren't laws at all. However, under a more liberal interpretation, those that have the fewest exceptions may still be considered to have a law-like status.

Ironically, the best candidates for biological laws aren't called laws, while those proposed generalizations that typically are called laws in biology shouldn't be. An example of the former is natural selection – usually called a theory – while an example of the latter is provided by Mendel's laws of inheritance. Arguably, there are no exceptions to natural selection, though I'll expand on that statement shortly. In contrast, there are many exceptions to Mendel's laws. Mendel's first law, which we now interpret as the two copies of each of our chromosomes separating (or segregating) and going into different germ cells, doesn't apply in the world of organisms with wholly haploid life cycles, because there aren't two chromosomal copies in the first place. And Mendel's second law, which can be stated as the independence of the segregation occurring at one gene locus from the segregation happening at another, is even more restricted in its applicability. Not only is it inapplicable to haploid organisms, but in the diploid realm it's only true of pairs of genes that are on different chromosomes.

So, the 'laws' of inheritance have plenty of exceptions here on Earth. Consequently, proposing that they should necessarily apply to the entire biological universe would be a very risky hypothesis indeed, and not one that I would be prepared to advance. But maybe natural selection, though variously called a mechanism, a process, and a theory (as we've already seen), is a universal biological law. Let's consider this possibility in some detail.

All life-forms on Earth live in groups called populations, albeit the nature of these varies widely. In the case of some animals, there are social interactions among the members of a population, while in the case of other animals, plants, fungi, and microbes, there aren't – unless sociality were to be defined in an unusually broad way. In all sexually reproducing organisms, whatever kingdom they belong to, there is genetic exchange among members of a population; and in this situation there is heritable variation. Even in the case of an asexually reproducing organism, where a population is simply a cohabiting batch of clones, gene mutations occurring in one or more of these clones will mean that there is heritable variation, though its extent will be less than in the case of its sexual counterpart.

Wherever there is reproduction and heritable variation, there is natural selection. What this means is that there is a *law* – as well as a mechanism and a theory – of natural selection. And here it is. All populations of life-forms are subject to natural selection. On Earth, it can be argued that there are no exceptions to this whatsoever. Alternatively, it can be argued that there are a few exceptions because there may be some small populations of some organisms that have *no* heritable variation (as opposed to just a very little variation, in which case natural selection would still happen).

Should we expect this law to apply also to extraterrestrial life? I don't see why not. Variation is omnipresent. It applies to galaxies, stars, planets, and life-forms. The biological realm on Earth stands out from its non-biological counterpart because not only is it characterized by variation but this variation is strongly linked with both reproduction and inheritance. This powerful three-way mix *is* natural selection. And as we saw at the outset, it is intrinsically bound up with our definition of life. It is likely to apply throughout the biological universe.

Strangely, it's possible to imagine a planet on which this *law* of natural selection applies, but Darwin's *theory* of natural selection does not. Recall that his theory can be stated in the form: natural selection is the main but not the only driver of evolution. If there is a planet on which the evolution of life features the inheritance of acquired characteristics – the evolutionary mechanism proposed unsuccessfully by Lamarck for Earthly life – there would be a combination of this mechanism and natural selection at work, and which of them would be responsible for the majority of evolutionary change would not be predictable without further information about the system concerned.

## Basic Units of Life

Our conclusion so far is that in terms of *processes* all of the biospheres that make up the biological universe may well be characterized by natural selection. But what about their *materials*? Might these be very different? After all, natural selection is defined in terms of entities (and groups of them) that exhibit reproduction, variation, and inheritance, *not* in terms of the construction of these entities. Despite that fact, there are two arguments in favour of the King Carbon hypothesis of Chapter 13 – the view that most if not all of the biological universe is based on organic molecules. One is the argument from chemistry: we have no evidence suggesting that any other element is capable of spawning anything that might be described as metabolism. The other is the Copernican biology approach, in which we acknowledge that not only is our physical location not special, but neither is our chemical composition.

According to the mediocrity principle – a close relation to the Copernican principle – if an item is drawn at random from a heterogeneous mixture of items belonging to several classes, some common and some rare, the item selected is more likely to be a representative of a common class than a rare one. This can be thought of in terms of the small ceramic balls we encountered in the previous chapter, except now instead of being spread over a visible grid they are hidden in a sack made of thick black canvas. Suppose the sack contains 100 balls, of which 90 are green, 9 are blue, and 1 is red. Providing they are thoroughly mixed, the most likely result of an experiment in which a person who cannot see into the sack picks one ball at random is a green ball. The least likely result is a red one.

Now replace the sack with the universe, and the balls with all instances of life; thus each ball is a biosphere. Suppose that there are 100 biospheres, 90 based on carbon, 9 based on silicon, and 1 based on some other element, say germanium. In such a universe, the chances that the only biosphere we know of is carbon-based are very high. But in an alternative universe where there are 90 germanium-based biospheres, 9 silicon ones, and just a single carbon one, the chances that the only biosphere we know is carbon-based are very low. Of course, the biological universe is likely to consist of many trillions of biospheres, not just a hundred; but this scaling up has no effect on the force of the argument.

In Chapter 13, we examined the building blocks of life not just at the molecular level but also at the next level up – the bags of molecules that we call cells. And we noted the near-universality of cellular construction of organisms on Earth. We considered the question of whether cells should be almost universal units of the construction of life throughout the Milky Way, and I hazarded the guess that membrane-enclosed cellular life is the norm, while in the vastness of the galaxy there will turn out to be some cases of life taking a different form. Here, we have two further jobs to do in relation to this Cell Centrality hypothesis: first, revisit the question of whether it has a sound basis; and second, ask whether it can be extended from one vast galaxy to an even vaster universe.

A sound basis for the hypothesis is the need that a life-form has for what can be called *manageable autonomy*. At the outset, part of our working definition of life was that it's characterized by a degree of buffering from whatever is going on outside. To maintain such an internal state, some sort of barrier is needed to separate 'organism' from 'environment'. The barrier needn't be – indeed mustn't be – impermeable. But it must have some control over what goes in and what comes out. So, *autonomy* argues for an outer membrane – as found in all true organisms on Earth – or at least some equivalent of it. *Manageability* argues for internal subdivision – also with a membrane or a close equivalent – in the case of any organisms that grow past a certain size, say about a metre across. As we've seen, the only exceptions to this on Earth are the acellular slime moulds; exceptions may be equally rare elsewhere.

To conclude, most of the life that makes up the biological universe is likely to be carbon-based. Perhaps even *all* of it takes this form, though given its vast extent, making such an assertion would be unwise. One level up, in constructional terms, most life-forms are likely to be cellular in the general sense of being composed of manageable units enclosed within membranes or some other forms of semipermeable barriers. But what about higher-still levels of construction, such as those that are involved in the bodies of large animals and plants? That's the realm we need to examine next.

## Big Bodies and Movement

On Earth, the animal and plant kingdoms represent the results of two quasi-independent experiments in multicellularity. They aren't the only ones; we saw earlier that some fungi and brown algae are the results of other such experiments. And they aren't truly independent, because plants and animals share a common ancestor – some long-extinct unicellular eukaryote – so it was always likely that both would retain the eukaryotic cell as a basis for their bodily construction. Indeed, their evolution was not independent for another, quite separate, reason: each provides a set of selective pressures that act on the other. The diverse morphologies of animals and plants can partly be attributed to such selection. For example, the forms of flowers and of pollinating insects have coevolved. So too have many other animal–plant relationships.

Plants and animals have reached far larger body sizes than the members of any other kingdom or domain. This is connected with a shared feature of these two kingdoms – their possession of supporting structures. But two other features *distinguish* the plant and animal kingdoms, even though there are occasional exceptions: presence versus absence of eating and moving. To a rough approximation, plants don't eat, and they don't move, whereas animals do both of these things. And it seems likely that these differences are connected with the multiple emergences of intelligence in one kingdom and its complete absence in the other.

We could perhaps say that of the two features – eating and moving – the former is logically prior to the latter. If an organism photosynthesizes, its energy is all around it, so there is no need to move. Strictly speaking, this is true only of the adult; the pollen and seeds that lead to the next generation must move, because the parents are rooted to a particular spot and their offspring cannot grow there. Whether they need to disperse a long or short distance depends both on the plant species and on the local environment. But this does not affect the generalization that adult plants do not move – in the sense of actively shifting their location. They can grow upward and outward, of course; they can bend towards light; and some floating pondweeds and airborne tumbleweeds can shift their locations, but only by passively responding to water currents or wind. There is no active movement.

In contrast, if an organism has to make its living by eating other organisms, movement is extremely useful, though not essential. Much here hangs on the definition of 'eating'. Fungi don't photosynthesize, but they don't eat either, if eating is deemed to mean taking in edible material via a mouth. Rather, they absorb carbon fixed by plants and other autotrophs from a substratum into which they are embedded or onto which they are encrusted. Like plants, they have a mobile stage in the life cycle – fungal spores – but also like plants, adult fungi do not actively shift their positions.

Although animals started off with an ecological *modus operandi* broadly similar to this, with ancient sessile sponges extracting both small organisms and organic detritus from seawater, they subsequently took an evolutionary route towards active movement. The embarking on this route can be regarded as the supreme defining feature of the animal kingdom. For sure, some groups of animals have backtracked to a sessile mode of life, but they are the exceptions rather than the rule. For the most part, the animal kingdom is characterized by organisms that are capable of active movement on various scales, ranging from small invertebrates that might move no more than the span of a human hand in their lifetime to migratory birds that move from one continent to another.

Because of the link between mobility and intelligence, we will now concentrate on Earthly animals and their extraterrestrial counterparts, as opposed to plants. However, it's worth pausing at this point to ask if 'counterparts' is the right word. I briefly mentioned John Wyndham's triffids in Chapter 5, these being hypothetical alien creatures with some features of plants (leaves) and some of animals (legs). They photosynthesize but are also carnivorous, like a Venus flytrap. However, unlike a flytrap, they can move around. There seems no good reason why such plant–animal life-forms should not exist somewhere, given the likely size of the biological universe. What follows may apply to them as well as to organisms that sit only on the animal side of the animal/plant fence.

The link between mobility and intelligence has been advocated, and explored, by several authors – including Matt Wilkinson, author of the beautifully titled 2016 book *Restless Creatures*. Here, we will focus on the issue of the directionality of movement. While the most basal animals (sponges) don't move at all as adults, some early-branching animals move, but in multiple directions. This includes various jelly-like forms – comb-jellies, box-jellies, and 'true' jellyfish. Watch a jellyfish swimming and you'll see that there is no 'forward' or 'backward', but rather an array of possible directions. This movement is powered by muscles, and these are activated by nerves, but there is no brain, no head, no front, and no rear.

One of the most important innovations in the history of the animal kingdom on Earth was the evolution, from this starting point, of a bilaterally symmetrical body-form that had a head end, a tail end, a dorsal surface, a ventral surface, a right side, and a left side. Welcome to the world of the worm. From the earliest worm-like beginnings in the pre-Cambrian oceans evolved the main group of animals, represented today by more than a million species: the Bilateria. Bilaterally symmetrical animals have a head end and a tail end. They have forward and reverse as their two principal directions of movement. Given the predominance of forward movement, natural selection favoured the concentration of sense organs, and of the brain that processes their inputs, at the head end of the animal. In this way, the first steps were taken along the evolutionary path of cephalization – elaboration of the head and brain.

As with other evolutionary trends, this particular route was just one out of many possible routes, rather than a compulsory ladder that all bilaterians had to climb. The most common path from proto-worm with proto-brain was towards moderately enhanced brains. Of course, given the overwhelming predominance of insect lineages in the animal kingdom, 'most common' tends to be another way of saying 'what the insects do'. However, elsewhere than among the insects, what we might call creeping cephalization is also the norm. But there are exceptions to this trend, and they are of two kinds: de-cephalization towards a smaller brain or even loss of a brain; and mega-cephalization towards high levels of intelligence. Echinoderms, bivalve molluscs, and parasitic barnacles provide examples of the former. Apes and octopuses provide examples of the latter. To what extent intelligent animals elsewhere might be similar to these Earthly forms is an open question.

## A Huge Hypothesis

Now let's draw together several possible generalizations about the biological universe into one Huge hypothesis. It's huge in several ways. First, its spatial scope is huge, covering billions of light years. Second, its temporal scope is huge, covering billions of Earth years. Third, it's huge in that it's a compound of many component hypotheses. Finally, it's hugely risky. Take it with the proverbial pinch of salt; indeed, an entire salt cellar's worth might be more appropriate in this case. However, there is one respect in which it is not huge – number of pages (less than two). Here it is (see also the even shorter version given in Figure 19):

The biological universe came into being 'a long time ago in a galaxy far, far away'. To go beyond this famous *Star Wars* opener: it came into being more than 10 billion years ago, when the first life-forms evolved from non-biological organic matter on a planet orbiting a star in a galaxy that, from a probabilistic perspective, was almost certainly not our own or even a member of our galaxy cluster. After the first galaxy 'went green', many others followed. And within each green galaxy, the number of green dots (planetary systems with life) gradually increased. Perhaps by now the numbers of green galaxies, and the densities of green dots within them, have stabilized, with the rates of generation and loss of biospheres being equal, but perhaps the

1. The first life evolved in the universe more than 10 billion years ago.

2. This original life evolved on a planet in a galaxy distant from our own Milky Way.

3. Since then, the number of galaxies with life has increased, as has the number of inhabited planets per galaxy.

4. The number of planets in the universe with life at present is many trillions.

5. Some planetary systems have more than one inhabited planet.

6. The vast majority of life in the universe (perhaps all of it) is carbon-based.

7. The vast majority of life in the universe is constructed of cells or cell-like units.

8. Most inhabited planets only have microbial life (single cells and small aggregations of cells).

9. Very many (but proportionally very few) inhabited planets have multicellular life.

10. Some planets with multicellular life have equivalents of our own animal and plant kingdoms, others don't.

11. Wherever mobility of large multicellular creatures has evolved it has gone hand-in-hand with the evolution of muscles, a nervous system, and a brain.

12. Across all the planets on which such evolution has happened, on some the most intelligent creature is a worm, while on others it is something like ET. All intermediate situations exist.

**Figure 19** Short version of the Huge hypothesis regarding the evolution of life in the universe, in which the twelve key points are listed. The first three refer to the past, the other nine to the present. The future is not covered. It's particularly important here to pay attention to the use of the following words: all, most, more, many, few, some, others, and majority.

numbers are still increasing. Either way, there are currently many trillions of green dots scattered across the observable universe, and perhaps many trillions more beyond it, in the realm of 'all the light we cannot see'.

If each green dot is a planetary system with life, there are some cases (many in number, few as a fraction) in which more than one planet is inhabited. Double- or multiple-life systems could arise in three ways: first, because life evolved independently on both/all of the planets concerned; second, because of the chance – although it is a very slim one – of spores travelling successfully between planets in cases where the distance between them is short, as in the TRAPPIST-1 system; third, because intelligent life may have colonized nearby planets, as may be the case in the mid-term future with humans on Mars. However, despite these possibilities, the commoner situation is likely to be just a single inhabited planet per system (with zero inhabited planets being commoner again).

In the vast majority of cases, life is carbon-based, though there may be very occasional exceptions. Carbon-based life is typically cellular, with organisms consisting of membrane-bound packages of material that have a degree of autonomy from the environment; but again, there may be exceptions. Many cellular-life planets have nothing more than microbes, in the form of organisms that are unicellular throughout their life cycles. In some cases, this is because the planets don't last long enough for multicellularity to evolve, perhaps because they orbit short-lived stars or are blown apart by collision with other large rocky bodies. In other cases, it is because the relevant type of heritable variation never occurred.

On some planets (again numerically many but proportionally few), larger body sizes evolved. In the majority of cases these took the form of multicellular creatures, though in a few cases alternative routes to large size were taken, so that there may be some planets dominated by large amorphous creatures somewhat similar to Earthly slime moulds. Some planets with multicellular life have broad equivalents to the animal and plant kingdoms on Earth, while others do not. Some of these other planets include in their biota plant–animal melanges such as triffids. Where mobility of large life-forms occurs, then regardless of whether it is found in animals or in triffid-like creatures, it is powered by contractile tissues (muscles) that are activated by signalling tissues (nerves).

On the majority of planets in which large organisms became able to shift their location by active movement, bilateral symmetry evolved, resulting in the origin of a head and a brain. On most of these planets some lineages evolved in the direction of increasing cephalization, with the result that the intelligence of the life-forms belonging to these lineages increased to various degrees. There are some planets on which the cleverest creature is a worm; some on which the highest level of intelligence is equivalent to an Earthly octopus or crow; and some on which broadcasting civilizations considerably more advanced than our own have evolved.

That's the end of the Huge hypothesis for now. We'll deal further with the intelligent universe (just touched on in the previous paragraph) in Chapter 20. But before leaving this one I'd like to point out that in outlining the Huge hypothesis I have used direct statements, apparently of fact (for example, 'many planets *have* animal life'), rather than using a less direct verb-form such as the conditional (for example 'many planets *would have* animal life'). Please note that I have only been able to do this because the whole section is labelled 'hypothesis'. Covered by this implicit overall doubt, simple direct statements work best, and the doubt does not have to be endlessly reiterated. Darwin's bulldog, Thomas Henry Huxley, excused himself for doing something similar in a long letter (dated 7 June 1852) to his colleague Albany Hancock: 'I have said many things in the indicative mood which would, I am aware, have been more becoming put in the potential – but the former was more concise & on that ground pray forgive it.' Like Huxley, I beg the reader's forgiveness in this respect.

# 20 THE INTELLIGENT UNIVERSE

## The Universe and its Children

There are three possible interpretations of the phrase 'the intelligent universe'. First, the universe itself is intelligent. It might be some sort of knowingly self-sustaining Gaia-like entity. But it probably isn't. Second, the universe might have been made by an intelligent designer – a supreme being of some sort. But it probably wasn't. This leaves us with only the third possible interpretation of 'the intelligent universe', in which the intelligence applies to life-forms scattered here and there, dotted about the vastness of space that the universe spans. We might think of these as the intelligent children of the universe, though of course they are part of it, not distinct from it. Indeed, the intelligent universe – the collection of all intelligent beings everywhere – is a subset of the biological universe, just as that in turn is a subset of the physical universe.

But there's a caveat here. What about AI (artificial intelligence)? It's a moot point whether any of our machines are yet intelligent enough to truly merit that label, though no doubt they will get there eventually. Perhaps the machines associated with ultra-intelligent aliens are already there. In this case, the intelligent universe and the biological universe have a different relationship – they are overlapping sets. Having made this point, let's focus on *intelligent living beings* across the universe, not intelligent machines. And let's ignore the advanced organism–machine hybrids of science fiction, even though entities of this type probably exist somewhere, albeit not in the form of time-travelling Terminators.

Towards the end of the previous chapter's Huge hypothesis, we envisaged intelligent life existing on multiple planets across the universe. We have taken such life to be numerically common but proportionally rare. In other words, there are many planets in the universe with intelligent life, but they comprise only a tiny fraction of the 'gazillions' of planets that exist, and indeed only a small fraction of those planets that have life. However, we also noted that intelligence is a continuum, not a binary variable: there are *degrees* of intelligence. It's a notoriously difficult thing to measure, but it starts at zero. Its upper bound is a complete unknown, but a safe bet is that the highest intelligence in the universe – however it's measured – is way beyond those Einsteinian levels of 200+ on our very imperfect IQ scale.

Given the continuous variation in intelligence, we can only consider a 'number of planets with intelligent life' if a particular *threshold level* of intelligence is used. It seems reasonable to expect the number of planets with life that is more intelligent than 'level X' to decrease as X increases. This is in a sense parallel to saying that a threshold 'large' body size, and the number of planets with life that big, are negatively correlated. There are probably more planets with microbes than with beetles, and more with beetles than with dinosaurs. The only complication in this parallel is that very intelligent creatures can colonize other planets, whereas very large but intellectually challenged ones, like dinosaurs, can't.

Let's define four threshold levels of intelligence. If we take organisms without nervous systems, whether bacteria, roses, or sponges, as having zero intelligence, then we can think of animals with a small brain, and with only slight behavioural flexibility, as having crossed the first threshold, and being at level 1. Earthworms and snails are good examples. Animals that can use tools, and indeed plan their use of tools, whether an octopus carrying coconut shells to a locality where it wants to use them as a hiding place, or a crow using one tool to acquire another, have crossed the second threshold, and are at level 2. Animals that have begun to investigate the abstract nature of things, and to keep written records of their investigations, have crossed the third threshold, and are at level 3. The ancient Greeks crossed this threshold between two and three millennia ago. Finally, from a pragmatic point of view, in terms of possible communication between intelligences on different planets, we should recognize as our fourth threshold the achieving of a civilization with a technology that includes the use of radio signals and other means of interstellar communication, such as lasers. We can say that the beings who created such a civilization have crossed the next threshold and thus have reached intelligence level 4.

Even without any data from beyond the Earth, it's hard to believe that the number of planets whose evolutionary processes have crossed these four respective thresholds would go upward rather than downward. There are almost certainly more planets where the maximum level of intelligence is zero than there are planets where intelligence has begun (crossed the first threshold). And the direction of change in the number of planets will be the same for any two consecutive levels: there are always more planets on which intelligence has crossed threshold X than threshold X+1. This particular wording avoids the issue of possible planetary colonization by the level 4s, because it involves the number of evolutionary crossings of the various thresholds on the relevant 'home planets', not what happens after such crossings.

Being able to quantify the declining numbers of planets with increasing levels of intelligence is impossible, given our lack of information from beyond the Earth. But perhaps the comparative timespans of the different forms of life on Earth give us some clues. Non-intelligent life has existed here for some 4 billion years. Simple worm-like animals with small brains have existed for only about 0.6 billion (or 600 million). Animals that can plan their use of tools have probably been around for about half of that time – the earliest fossil octopus dates from just under 300 mya, while birds and mammals both originated later than that. There's then another gap of about 300 million years, because the crossings of the third and fourth thresholds were simultaneous and happened at 0 mya, if we measure things only to the nearest million years.

Using this sort of information from the Earth, we can guestimate that the fraction of planets with life that have crossed intelligence threshold 1 is about 15% (0.6 billion divided by 4 billion), while the fraction that have crossed threshold 2 is about half of that; and the fractions that have crossed thresholds 3 and 4 are minuscule. Of course, there will be all sorts of reasons why these Earth-based guestimates are wrong, one of which is that Earth is probably only about halfway through its potential evolutionary history. But even so, they provide a better starting point than no guestimates at all. And it's not necessarily true that they're only based on a sample size of one, for the following reason.

For sure, we still only have evidence of life on one planet. However, we can study every lineage of life on our planet, and to some extent each of these is an independent experiment in finding an evolutionary direction. It's possible to lose a brain as well as to gain one, as the echinoderms show us. It's possible to get smaller as well as to get bigger, as miniature mammals like bumblebee bats make clear. It's possible to lose adult mobility, as in the barnacles. And it's possible to lose sense organs, as exemplified by blind cave fish. Evolution is not, and never has been, a one-way process. Every lineage takes its own route, albeit with the baggage of its past and the selective pressures generated by its cousins as constant reminders of the difference between the quasi-independence that we actually see and the 'complete independence' that's just a theoretical construct. Examining different paths to intelligence on Earth is helpful in that these paths *may* apply in a wider context.

## Milestones to Intelligence

On Earth, intelligence is associated with certain things. Most obviously, it's associated with a brain. But it's also associated with certain other structures, senses, and behaviours. The brain itself is probably a result of directionality of movement, as we saw in Chapter 19. And forward movement is also associated with eyes. Although box-jellies have surprisingly complex eyes, most animals that can see the world are bilaterally symmetrical and have eyes at their anterior ends. Intelligence has never arisen in a lineage of blind animals. However, we have to be careful with such reasoning, because intelligence has never arisen in an animal that lacks an anus – yet there is no direct causal relationship between anus and brain, both are simply typical features of a bilaterally symmetrical body-plan. Also, most animals have eyes, so the fact that the few intelligent animals have eyes is not necessarily informative. And you can't pass the mirror test if you can't see.

However, the situation is altogether different when it comes to arms – broadly defined as highly mobile appendages whose function is not restricted to propulsion. In Chapter 8, we regarded arms as 'helpful' to the evolution of intelligence, yet neither 'necessary' nor 'sufficient'. Let's now probe further into this issue.

The most intelligent vertebrates on Earth – apes (almost 30 species) – and the most intelligent invertebrates – octopuses (about 300 species) – both have arms. Not only that, but apes have fingers and octopuses have suckers. Is the link between the manual dexterity that applies in both cases and the evolution of intelligence a coincidence? I think not. Admittedly, both apes and octopuses belong to groups in which intelligence is already quite high. But then again, the possession of arms is common to these broader groups too. All primates have arms with hands, and digits that can be called fingers; all cephalopods have tentacles, and all of the coleoid cephalopods – the group to which octopus, squid, and cuttlefish belong – have suckers. In both cases – primates and cephalopods – the evolutionary sister-groups lack arms/hands/fingers and tentacles/suckers, and are characterized by lower intelligence. So, one obvious hypothesis is that intelligence and manual dexterity are causally related. However, we need to ask to what degree this hypothesis is plausible. And one component of this question is how common it is for animals to have manipulative appendages that can be called arms or tentacles, outside of the primates and the cephalopods.

As ever, the devil is in the detail. Arms with hands that have fingers and opposable thumbs are rare outside the primates. Pandas have a 'pseudo-thumb' that is opposable and enables them to grasp things – like bamboo stalks. Sloths have hand-like front feet, but no opposable thumbs. Kangaroos can use their front paws in an almost hand-like (and fist-like) way, but again there is no opposable thumb.

Structures called tentacles are found in many invertebrate groups, but they are usually much less elaborate than in ceph-alopods, and to emphasize the difference many are called cirri (singular cirrus). These are found, for example, in nemertean worms and in the sessile echinoderms called sea lilies. Tiny tentacles that are not usually called cirri are found in other invertebrates, notably in moss animals (bryozoans), where they are part of a feeding structure called a lophophore. More sub-stantial tentacles are found in jellyfish, but these are primarily stinging structures, not manipulative ones; and in any event they could hardly be associated with intelligence, since jellyfish have no brain.

Given the rarity of arms/hands/fingers/thumbs on the one hand and large flexible tentacles with suckers on the other, their association with intelligence seems unlikely to be coincidental. However, manual dexterity is not a prerequisite for a *reasonably* high level of intelligence (level 2), since it is absent in crows and dolphins. And in general, we should be very hesitant to label anything – with the exception of a brain – as being such a prerequisite. Rather, a better way to describe the situation in the animal kingdom on Earth would be that there are several factors that increase the likelihood of intelligence evolving. These include: directional movement, bilateral symmetry, manual dexterity, sociality, and language. Whether these same factors will turn out to apply elsewhere than Earth is an inter-esting question.

One way to consider this issue is represented – in the form of milestones – in Figure 20. Here, we make just two assumptions

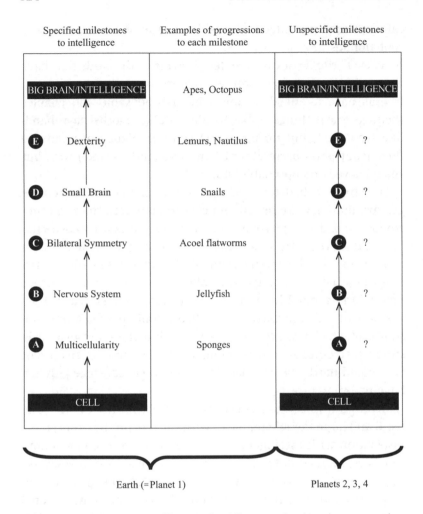

**Figure 20** Milestones to intelligence. Five milestones that have been crossed on the evolutionary journeys to the most intelligent vertebrates (apes) and invertebrates (octopuses) on Earth (left), together with examples of animals that have reached each milestone but not (so far) beyond it (centre). For other planets with life (right), we can ask the question of what the appropriate milestones (labelled A to E) are, and whether they might be similar to those of Earth.

about life on three other randomly chosen inhabited planets: first, that it is carbon-based; and second, that it starts off in the form of a creature that consists of a single unit that is quasi-autonomous from the environment – a cell, or something like it.

According to the Huge hypothesis of the previous chapter, most but not all life is of this kind. Having made these assumptions, we can take a series of five milestones to intelligence on Earth (left panel of Figure 20), and look at particular terrestrial animals as exemplifying those milestones (centre). Then we can denude the series of milestones of any detail and depict it in blank form (right) for our other three planets (the number is arbitrary), and ask the question: what are the milestones (A to E) in these extraterrestrial cases? Are they likely to be similar to those of Earth, or entirely different? Naturally, this is an open question, though I can't help but feel that at least one out of our three randomly chosen planets would have milestones rather similar to ours.

## Is There Enough Time?

When evolution begins on any particular planet, it has no way of 'knowing' how much time is available to it. As we've noted before, and as captured in Richard Dawkins's book-title *The Blind Watchmaker*, natural selection is a process that cannot look into the future. It deals very much with the here-and-now. And there is no reason for it to be any different in this respect on another planet. So, once the first metabolizing, reproducing proto-cells are available for natural selection to work on, the process will kick off regardless of whether it has a few million years or a few billion years to run. Indeed, although its run-time is partly pre-set by the spectral class of the relevant star, a potentially long run-time might be drastically shortened by planetary collisions or other catastrophic events.

If we leave catastrophes aside, the time available for evolution will depend on two factors: the sun that is at the centre of the planetary system concerned, and the time that elapses on a particular habitable planet before life originates. If $T$ is the maximum potential lifespan of the sun and planet, and $t_1$ is the time from planet formation to the origin of life, then the maximum period for evolution, $t_2$, is simply $T - t_1$. On Earth, $T$ is thought to be about 10 billion years, whereas $t_1$ was about

0.5 billion. Therefore, $t_2$ on Earth (as a maximum, recall) is about 9.5 billion. Other stars that are similar to our Sun will have similar values of $T$. Whether habitable planets within their systems have $t_1$ values similar to that on Earth we have no idea, but values both longer and shorter are possible, and as usual it's best to assume that Earth is fairly middling, rather than constituting an extreme value or – worse still – representing some sort of outlier that is off the usual scale. And it follows that this argument applies to $t_2$ as well.

Let's assume that the distribution of values for $t_1$ on planets orbiting Sun-like stars is from about 100 million to 1 billion Earth years, adopting our familiar principle that we can't guestimate things to better than an order of magnitude. The distribution might be normal; or it might be skewed, given that there is a cut-off point at zero years. The distribution of $t_2$ values – the maximum times available for evolution – is then from 9.0 to 9.9 billion years, assuming similar stellar lifespans and ignoring the gradual brightening of stars over their main-sequence lifespans that we noted earlier. For stars that are more massive and more luminous than the Sun, the distribution of $t_2$ values is shifted downward, while for red dwarf systems it is shifted upward. Since there are far more red dwarf systems than all the others put together, the time available for evolution to produce results will typically be much longer than on Earth – tidal-locking issues permitting.

Unfortunately, there's another variable we need to take into account here that's a complete unknown: the speed of evolution on another planet. But again we can take a Copernican approach and assume that Earth isn't special. If this is true, then our evolutionary process here proceeds at a middling rate in terms of evolution in the biological universe as a whole. That helps a bit, but it doesn't help us to come up with an idea of how great the variation around such middling values is. It's likely to depend on many things, one of which is the generation time of the life-forms concerned. We know that on Earth evolution happens much more quickly when the organism that's evolving is a bacterium, with a generation time measured in minutes, than when it's a large animal or plant, with a generation time measured in decades. In an interstellar context, generation times may be loosely linked to planetary rotational and/or orbital periods, which vary widely, as we've seen.

What all this suggests is that there will be some habitable planets on which there is not enough evolutionary time for intelligence to evolve, and others on which there is more than enough such time. Let's now feed this provisional conclusion into our Huge hypothesis.

## Appendix to the Huge Hypothesis

To save you flicking back through the pages, here's how far we got, at the end of the relevant section of Chapter 19, in relation to the distribution of intelligence through the biological universe: 'There are some planets on which the cleverest creature is a worm; some on which the highest level of intelligence is equivalent to an Earthly octopus or crow; and some on which broadcasting civilizations considerably more advanced than our own have evolved.'

Can we go any further than this? Well, yes, we can try to imagine what 'considerably more advanced than our own' might mean. The background to this is the advance of human civilization, society, culture, science, and technology over time; and in particular the contrast between the speed of this advance and the speed with which evolution produced the brain that made it all possible. From the origin of life on Earth to the origin of a brain of about 1350 cubic centimetres in the genus *Homo* took about four billion years. From the origin of our large brain to the start of agriculture took from a bit less than a quarter of a million years ago until sometime between 20,000 and 10,000 years ago – say a span of 200,000 years. From agriculture to cities to the beginnings of science in ancient Greece took from (say) 15,000 until about 3000 years ago – a span of about 12,000 years. From the start of science to the theory of relativity, radio communication, and space travel took less than 3000 years. It's easy to perceive the general trend here. Everything is speeding up – though naturally there are always blips in a general trend, the Dark Ages of early medieval Europe being one of them.

But is this general acceleration of the rate of advance likely to be true of nascent level-4 (radio-defined) intelligence elsewhere? I think there are good reasons for suspecting that the answer is 'yes'. Cultural inheritance is likely to supplement biological inheritance wherever intelligence arises. And the building of technology on itself is a sort of runaway process – a positive-feedback system in which each advance helps to usher in others. If this is generally true, or even if it isn't and on some planets the rate of progress of civilizations is significantly slower than it has been on Earth, then the overwhelmingly important factor in determining how far intelligence might go is the available time. The single most important factor in determining this is the spectral class of the sun that life finds itself orbiting. And that's just the luck of the draw.

Let's now make the necessary addition to the Huge hypothesis so that it includes reference to the high end of the intelligence scale and the things that go with it, such as science and technology. As in Chapter 19, I'll use a direct form of words, since we're already covered for uncertainty by the heading of 'hypothesis':

Planets on which radio-level intelligence has evolved constitute only a tiny fraction of those on which life in general has evolved. Yet because of the vastness of the universe, and perhaps also because of planetary colonization, there are many planets with such life-forms in the universe right now. As our earlier experiments with the Drake equation showed, the number of such planets in the Milky Way could be as low as one (us), though it's probably higher – up to about a quarter of a million. But in the universe as a whole, even restricting our attention – as we must – to the observable universe, the number is huge. Even with only one intelligent civilization per galaxy there are at least a trillion of them, some new, some middle-aged, some old – I call this the Trillion Intelligences hypothesis.

With regard to the age and stage of intelligent civilizations, we humans are only about a century into the radio age. There are almost certainly other life-forms that are a thousand, a million, and perhaps even a billion, years past this threshold for level-4 intelligence. This means that some will by now have passed the thresholds for levels 5, 6, etc. – even though we don't yet know what these thresholds are. There must be civilizations 'out there' that truly make us look primitive. They may be forever beyond our reach (and we beyond theirs) due to a combination of Einstein's universal speed limit and the accelerating expansion of the universe that's taking us all inexorably towards the big freeze. Or we might have contact with them in the next decade by radio, or even meet them via a 'wormhole', if such things turn out to exist. What would first contact be like?

## First Contact

This phrase is used both for the first meeting of two different human cultures in the past and for the first meeting between the human species and intelligent extraterrestrial life-forms in the future. In the former case, which we know quite a lot about since it has happened on many occasions, the outcome is usually bad for one of the cultures concerned. In particular, when one culture has at its disposal a more advanced technology, the other culture typically suffers, up to the point where it is effectively destroyed. That's not to say that the culture becomes completely extinct, but rather that it's reduced to a vestige of what it was. Examples of this are provided by the arrival of Europeans in the Americas and in Australia, with dire consequences for the native peoples of those lands. Differences in technology are not the only factor in these cases – introduction of disease is another – but it is a decisive one. And while many aspects of technology are involved, the possession of more advanced weaponry is particularly important in determining the outcome.

Will our first contact, as a species, with intelligent alien life-forms from beyond the solar system have similarly adverse consequences, and if so for whom – us or them? To answer the second part of the question first: we humans are likely to have the inferior technology if first contact happens anytime soon, given that we are a mere century into the radio age, while the other participant could be vastly beyond that point. It's *possible* that we could receive a message tomorrow from a civilization about 50 light years away that was where we were in technological terms in 1974, when the Arecibo message was sent into space – in which case we and they may be at a very similar stage of technology. But it's extremely unlikely. If an alien spacecraft arrives on Earth tomorrow, then for sure we are the less technologically advanced culture. The odds are heavily stacked against us in this respect; we only stand a chance of having the more advanced technology if first contact is many millennia into the future.

This means that potentially we have much to fear from such contact, as it might indeed take the form of the eradication of human culture, if the other life-form is hostile. But we also have much to hope for, if the other life-form is friendly: the subsequent acceleration of the rate of increase in human knowledge might be phenomenal. It's this combination of great hope and great fear that makes it hard for us to know how to anticipate such an event. Two comments made by Stephen Hawking are pertinent here. In 2010, he said in a BBC interview: 'If aliens visit us, the outcome would be much as when Columbus landed in America, which didn't turn out well for the Native Americans.' Five years later, when participating in the launch of the Breakthrough Initiatives, he said: 'In an infinite universe, there must be other life. There is no bigger question. It is time to commit to finding the answer.' (This wording is from an account in *Nature* by Zeeya Merali; be warned that various versions of this quote exist.) Perhaps humanity will be a victim of its endeavour to answer the biggest questions; but then again, perhaps not.

The key issue here is how the probability of a culture being 'friendly' (positive intentions towards others, no interest in acquiring their territory, merely visiting) or 'hostile' (negative intentions, acquisitive, invasive) depends on how advanced it is. The two extreme possibilities are nicely illustrated by the British cult sci-fi series *Doctor Who*, in which the eponymous Doctor, a highly advanced maverick time-lord, belongs in the friendly category, while the infamous Daleks, whose favourite word is 'exterminate', belong in the hostile one. But what about *real* aliens? Perhaps the best clue to those is to look at our own history.

Has our own position, in terms of friendliness versus hostility to other cultures and other species on Earth, changed as we have advanced in technology? To begin with, the answer was probably either 'no' or 'yes, but not in a good way'. However, there are some promising recent signs. The conservation movement, whose roots go back several hundred years, mushroomed in the twentieth century. And the principle of planetary protection, adopted by NASA and the equivalent organizations of other countries, is a commendable approach, representing a desire not to contaminate possible life-supporting planets or moons with terrestrial microbes.

There are two problems in adopting an overall optimistic approach based on such recent positive observations on humanity. The first is that we're not far up the scale of technological advance, so we can't know whether moves towards respect for other cultures, species, and planetary homes are part of a trend that will continue linearly as human technology advances, or whether there might be more complex patterns of change. Nor do we know whether such a long-term linear trend between advancing technology and an increasingly enlightened approach to others – even if one exists in humanity – will characterize intelligent life-forms elsewhere. The second problem is that we cannot be sure of the motives of aliens making radio contact or coming to visit us. Our own exploration of space is driven by scientific interest, not a desperate quest for survival, as in the case of a farmer emigrating to the USA from famine-ravaged Ireland in the 1850s. People adopt a more enlightened approach to others when their own survival is not under threat. This may also be true of aliens, so we should hope to meet those whose home planet is not under threat, and who are thus not in need of acquiring ours.

Not knowing what to expect in terms of hostility versus friendliness from intelligent extraterrestrials is not the same as expecting the worst. Our uncertainty should urge a degree of caution to be sure, but not a retreat from activities such as SETI. In any event, extraterrestrials may contact us before we contact them – whether by radio messages or arrival. Given that fact, we need to be as informed as possible. So the growth of interest in astrobiology is to be welcomed, not seen as a threat. And right now, the greatest enemy of the scientific endeavour – and indeed of enlightened humanity in general – is not in deep space, but right here on our home planet.

## The Present Era: Earth

Today, we are privileged to live in an age of reason, an age of science. Accounts of when this era began vary widely. My personal preference is to think of its roots extending back to the year 1543, which saw the publication of Copernicus's magnum opus *On the Revolutions of the Celestial Spheres*. But any attempt to see history as a sequence of well-defined ages or eras is fatally flawed. Not only are transitions between eras often not clearly defined, but each era is heterogeneous, especially when we think across the entire world, rather than restricting our attention to a single part of it, such as a particular country or continent. Unfortunately, in the era of conservation, poaching rare animals for the fictitious healing power of their ground-up body parts is still rife in some parts of the world. And although state-sponsored murder of heretics by burning at the stake is thankfully a thing of the past in Europe – where it used to be common – there are still more than a dozen countries in the Middle East, Africa, and Asia where the penalty for atheism is death. The age of reason – and of free thought and free speech – has regrettably not yet permeated the whole world. With luck, it eventually will, but of this we have no guarantee whatsoever.

Einstein once said that science is 'the most precious thing we have'. While other things are precious too in a general sense (love, compassion, and forgiveness are high on the list), in the sense of understanding the universe Einstein was absolutely right. The scientific endeavour, based on insight, hypotheses, reason, and evidence, is a jewel in humanity's crown. Our crown may be a small one by universal standards, but on Earth it is precious indeed.

In the near future, one of our most urgent tasks as a world-scattered, if not yet worldwide, civilization is to protect the scientific endeavour, and the reasoned, free-thinking approach that underlies it, from multiple threats. It matters not from where these come; we should take them all seriously. Hostility to science can come from advocates of intelligent design, and indeed from fundamentalists of any religion who are intolerant of free-thinking 'infidels' and believe that unquestioning faith is superior to reason. It can also come from political movements that subvert science to ideology, as the Soviet regime did with the field of genetics in the era of Lysenkoism. Even in present-day democratic regimes, which ought to know better, science denial can be a serious threat. A 'dark age of the future' is not impossible. Let's strive to avoid such a fate by championing in every way we can the pursuit of our science and the enhancement of our age of reason.

# BIBLIOGRAPHY

Anglada-Escudé, G. *et al.* (30 authors) 2016. A terrestrial planet candidate in a temperate orbit around Proxima Centauri. *Nature*, 536: 437–440.

Archibald, J. 2018. *The Tangled Tree: A Radical New History of Life*. Simon & Schuster, New York.

Arthur, W. 2004. *Biased Embryos and Evolution*. Cambridge University Press, Cambridge.

Arthur, W. 2017. *Life through Time and Space*. Harvard University Press, Cambridge, MA.

Barnes, R. 2017. Tidal locking of habitable exoplanets. *Celestial Mechanics and Dynamical Astronomy*, 129: 509–536.

Bluff, L.A., Weir, A.A.S., Rutz, C., Wimpenny, J.H. and Kacelnik, A. 2007. Tool-related cognition in New Caledonian crows. *Comparative Cognition and Behavior Reviews*, 2: 1–25.

Boal, J.G., Dunham, A.W., Williams, K.T. and Hanlon, R.T. 2000. Experimental evidence for spatial learning in octopuses (*Octopus bimaculoides*). *Journal of Comparative Psychology*, 114: 246–252.

Bobrovskiy, I., Hope, J.M., Ivantsov, A., Nettersheim, B.J., Hallmann, C. and Brocks, J.J. 2018. Ancient steroids establish the Ediacaran fossil *Dickinsonia* as one of the earliest animals. *Science*, 361: 1246–1249.

Caldeira, K. and Kasting, J. 1990. The lifespan of the biosphere revisited. *Nature*, 360: 721–723.

Charbonneau, D., Brown, T.M., Noyes, R.W. and Gilliland, R.L. 2002. Detection of an extrasolar planet atmosphere. *Astrophysics Journal*, 568: 377–384.

Chen, Z., Zhou, C. and Xiao, S. 2019. Death march of a segmented and trilobate bilaterian elucidates early animal evolution. *Nature*, 573: 412–415.

Cocconi, G. and Morrison, P. 1959. Searching for interstellar communications. *Nature*, 184: 844–846.

Cohen, J. and Stewart, I. 2002. *Evolving the Alien*. Ebury Press, London. (Later editions published under the title *What Does a Martian Look Like?*)

Conway Morris, S. and Peel, J.S. 1990. Articulated halkieriids from the Lower Cambrian of north Greenland. *Nature*, 345: 802–805.

Darwin, C. 1859. *On the Origin of Species by Means of Natural Selection, or the Preservation of Favoured Races in the Struggle for Life*. John Murray, London.

Darwin, C. 1881. *The Formation of Vegetable Mould through the Action of Worms, with Observations on their Habits*. John Murray, London.

Davies, P.C.W., Benner, S.A., Cleland, C.E., Lineweaver, C.H., McKay, C.P. and Wolfe-Simon, F. 2009. Signatures of a shadow biosphere. _Astrobiology_, 9: 241–249.

Dawkins, R. 1986. _The Blind Watchmaker_. Penguin, London.

De Wit, J. _et al._ (17 authors) 2018. Atmospheric reconnaissance of the habitable-zone Earth-sized planets orbiting TRAPPIST-1. _Nature Astronomy_, 2: 214–219.

Demory, B.-O. _et al._ (14 authors) 2013. Inference of inhomogeneous clouds in an exoplanet atmosphere. _Astrophysical Journal Letters_, 776: L25.

Doerr, A. 2007. Window of possibility. _Orion Magazine_, 25 June 2007. Available at: https://orionmagazine.org/article/window-of-possibility/ (accessed February 2020).

Doerr, A. 2014. _All the Light We Cannot See_. Fourth Estate, New York and London.

Drake, F. 1961. Discussion of Space Science Board, National Academy of Sciences conference on extraterrestrial intelligent life, November 1961. Green Bank, West Virginia.

Elton, C.S. 1927. _Animal Ecology_. Sidgwick & Jackson, London.

_Extrasolar Planets Encyclopaedia_. Available at: http://exoplanet.eu/ (accessed February 2020).

Finn, J.K., Tregenza, T. and Norman, M.D. 2009. Defensive tool use in a coconut-carrying octopus. _Current Biology_, 19: R1069–R1070.

Finney, J. 2015. _Water: A Very Short Introduction_. Oxford University Press, Oxford.

Fortey, R. 2000. _Trilobite! Eyewitness to Evolution_. Harper Collins, London.

Fraine, J. _et al._ (9 authors) 2014. Water vapour absorption in the clear atmosphere of a Neptune-sized exoplanet. _Nature_, 513: 526–529.

Fusco, G. and Minelli, A. 2019. _The Biology of Reproduction_. Cambridge University Press, Cambridge.

Gallup, G.G. 1970. Chimpanzees: self-recognition. _Science_, 167: 86–87.

Gillon, M. _et al._ (30 authors) 2017. Seven temperate terrestrial planets around the nearby ultracool dwarf star TRAPPIST-1. _Nature_, 542: 456–460.

González-Forero, M. and Gardner, A. 2018. Inference of ecological and social drivers of human brain-size evolution. _Nature_, 557: 554–557.

Gould, S.J. 1989. _Wonderful Life: The Burgess Shale and the Nature of History_. Norton, New York.

Grant, P.R. 1986. _Ecology and Evolution of Darwin's Finches_. Princeton University Press, Princeton, NJ.

Grant, P.R. and Grant, B.R. 2011. _How and Why Species Multiply: The Radiation of Darwin's Finches_. Princeton University Press, Princeton, NJ.

Hart, M.H. 1979. Habitable zones about main sequence stars. _Icarus_, 37: 351–357.

Hawking, S. 1988. _A Brief History of Time: From the Big Bang to Black Holes_. Bantam Books, New York.

Hoyle, F. 1957. _The Black Cloud_. Heinemann, London. (Reprinted as a Penguin Classic, 2010.)

Huang, S.-S. 1959. Occurrence of life in the universe. _American Scientist_, 47: 397–402.

Huxley, T.H. 1852. Letter to Albany Hancock, 7th June. Transcript available at: http://nhsn.ncl.ac.uk/wp-content/uploads/2017/09/Correspondence-AH-THH-Final-Edit.pdf (accessed February 2020).

Javaux, E.J. 2019. Challenges in evidencing the earliest traces of life. *Nature*, 572: 451–460.

Jenkins, J.M. *et al.* (29 authors) 2015. Discovery and validation of Kepler-452b: a 1.6$R_\oplus$ Super Earth exoplanet in the habitable zone of a G2 star. *The Astronomical Journal*, 150: 56 [19 pages].

Jonnson, K.I., Rabbow, E., Schill, R.O., Harms-Ringdahl, R. and Rettberg, P. 2008. Tardigrades survive exposure to space in low Earth orbit. *Current Biology*, 18: R729–R731.

Kasting, J. 2010. *How to Find a Habitable Planet*. Princeton University Press, Princeton, NJ.

Kasting, J., Whitmire, D.P. and Reynolds, R.T. 1993. Habitable zones around main sequence stars. *Icarus*, 101: 108–128.

Kimura, M. 1983. *The Neutral Theory of Molecular Evolution*. Cambridge University Press, Cambridge.

Kivelson, M.G., Khurana, K.K., Russell, C.T., Volwerk, M., Walker, R.J. and Zimmer, C. 2000. Galileo magnetometer measurements: a stronger case for a subsurface ocean at Europa. *Science*, 289: 1340–1343.

Lack, D. 1947. *Darwin's Finches*. Cambridge University Press, Cambridge.

Lewontin, R.C. 1974. *The Genetic Basis of Evolutionary Change*. Columbia University Press, New York.

Lubin, P. 2016. A roadmap to interstellar flight. *Journal of the British Interplanetary Society*, 69: 40–72.

Mather, J.A. and Anderson, R.C. 1999. Exploration, play and habituation in octopuses (*Octopus dofleini*). *Journal of Comparative Psychology*, 113: 333–338.

Maxwell, J.C. 1865. A dynamical theory of the electromagnetic field. *Philosophical Transactions of the Royal Society of London*, 155: 459–512.

Maxwell, J.C. 1873. *A Treatise on Electricity and Magnetism*, Volumes I and II. Clarendon Press, Oxford.

Merali, Z. 2015. News: search for extra-terrestrial intelligence gets a $100 million boost – Russian billionaire Yuri Milner announces most comprehensive hunt for alien life. *Nature*, 523: 392–393.

Milner, Y. *et al.* (32 authors) 2015. Open letter: 'Are we alone?' Available at: http://breakthroughinitiatives.org/arewealone (accessed February 2020).

*NASA Exoplanet Archive.* Available at: https://exoplanetarchive.ipac.caltech.edu (accessed February 2020).

Paine, R.T. 1966. Food web complexity and species diversity. *American Naturalist*, 100: 65–75.

Postberg, F. *et al.* (21 authors) 2018. Macromolecular organic compounds from the depths of Enceladus. *Nature*, 558: 564–568.

Pross, A. 2012. *What is Life? How Chemistry Becomes Biology*. Oxford University Press, Oxford.

Quintana, E.V. *et al.* (23 authors) 2014. An Earth-sized planet in the habitable zone of a cool star. *Science*, 344: 277–280.

Reiss, D. and Marino, L. 2001. Mirror self-recognition in the bottlenose dolphin: a case of cognitive convergence. *Proceedings of the National Academy of Sciences of the USA*, 98: 5937–5942.

Richardson, L.J., Deming, D., Horning, K., Seager, S. and Harrington, J. 2007. A spectrum of an extrasolar planet. *Nature*, 445: 892–895.

Sagan, C. 1980. *Cosmos*. Random House, New York.

Schrijver, K. 2018. *One of Ten Billion Earths: How We Learn About Our Planet's Past and Future from Distant Exoplanets*. Oxford University Press, Oxford.

Seager, S. 2010. *Exoplanet Atmospheres: Physical Processes*. Princeton University Press, Princeton, NJ, and Oxford.

Snellen, I.A.G., de Kok, R.J., le Poole, R., Brogi, M. and Birkby, J. 2013. Finding extraterrestrial life using ground-based high-dispersion spectroscopy. *The Astrophysical Journal*, 764: 1–6.

Stevenson, D.J. 1999. Life-sustaining planets in interstellar space? *Nature*, 400: 32.

Stott, R. 2004. *Darwin and the Barnacle*. Norton, New York.

Summers, M. and Trefil, J. 2017. *Exoplanets: Diamond Worlds, Super Earths, Pulsar Planets, and the New Search for Life Beyond Our Solar System*. Smithsonian Books, Washington, DC.

Tasker, E. 2017. *The Planet Factory: Exoplanets and the Search for a Second Earth*. Bloomsbury Sigma, London and New York.

Thompson, D'A.W. 1917. *On Growth and Form*. Cambridge University Press, Cambridge.

Tsiaras, A. *et al.* (11 authors) 2016. Detection of an atmosphere around the super-Earth 55 Cancri e. *The Astrophysical Journal*, 820: 99 [13 pages].

Tsiaras, A., Waldmann, I.P., Tinetti, G., Tennyson, J. and Yurchenko, S.N. 2019. Water vapour in the atmosphere of the habitable-zone eight-Earth-mass planet K2-18 b. *Nature Astronomy*, 3: 1086–1091.

Ward, P.D. and Brownlee, D. 2000. *Rare Earth: Why Complex Life is Rare in the Universe*. Copernicus Books, New York.

Watson, J.D. and Crick, F.H.C. 1953. A structure for deoxyribose nucleic acid. *Nature*, 171: 737–738.

Wegener, A. 1912. Die Entstehung der Kontinente. *Geologische Rundschau*, 3: 276–292. (English translation: von Huene, R. 2002. The origins of continents. *International Journal of Earth Science*, 91: S4–S17.)

Wilkinson, M. 2016. *Restless Creatures: The Story of Life in Ten Movements*. Basic Books, New York.

Winn, J.N. 2018. Shadows of other worlds. *Scientific American*, 318 (3): 26–33.

Woese, C.R., Kandler, O. and Wheelis, M.L. 1990. Towards a natural system of organisms: proposal for the domains Archaea, Bacteria, and Eucarya. *Proceedings of the National Academy of Sciences of the USA*, 87: 4576–4579.

Wyndham, J. 1951. *The Day of the Triffids*. Michael Joseph, London.

# INDEX

Page numbers in *italic* include illustrations.